Autotuning of PID Controllers

Cheng-Ching Yu

Autotuning of PID Controllers

A Relay Feedback Approach

2nd Edition

With 140 Figures

 Springer

Cheng-Ching Yu, PhD
Department of Chemical Engineering
National Taiwan University
1 Roosevelt Road
Section 4
Taipei 106-17
Taiwan

British Library Cataloguing in Publication Data
Yu, Cheng-Ching, 1956-
  Autotuning of PID controllers. - 2nd ed.
  1. PID controllers
  I. Title
  629.8

ISBN 978-1-84996-546-0   Printed on acid-free paper

© Springer-Verlag London Limited 2006
Softcover reprint of the hardcover 2nd edition 2006

First published 1999
Second edition 2006

MATLAB® is the registered trademark of The MathWorks, Inc., 3 Apple Hill Drive, Natick, MA 01760-2098, U.S.A. http://www.mathworks.com

Apart from any fair dealing for the purposes of research or private study, or criticism or review, as permitted under the Copyright, Designs and Patents Act 1988, this publication may only be reproduced, stored or transmitted, in any form or by any means, with the prior permission in writing of the publishers, or in the case of reprographic reproduction in accordance with the terms of licences issued by the Copyright Licensing Agency. Enquiries concerning reproduction outside those terms should be sent to the publishers.

The use of registered names, trademarks, etc. in this publication does not imply, even in the absence of a specific statement, that such names are exempt from the relevant laws and regulations and therefore free for general use.

The publisher makes no representation, express or implied, with regard to the accuracy of the information contained in this book and cannot accept any legal responsibility or liability for any errors or omissions that may be made.

9 8 7 6 5 4 3 2 1

Springer Science+Business Media
springeronline.com

To Patricia, Jessica, and Albert
獻給我的家人－鄭怜悧、余潔思、余肇偉

Preface

This edition is a major revision to the first edition. The revision is motivated by the new progress in relay feedback autotuning, as proposed by Bill Luyben, where the shape of the relay response can be utilized to identify likely model structure. Several new chapters have been added, notably the use of the shape-factor for autotuning and controller monitoring, incorporating autotuning in a multiple-model setup, dealing with an imperfect actuator. At the turn of the century, competitiveness in the global economy remains the same and the need for rapid and flexible manufacturing has become standard practice. This has given process control engineers an expanded role in process operation.

It has long been recognized that industrial control is one of the key technologies to make existing processes economically competitive. In theory, sophisticated control strategies–supervisory, adaptive, model predictive control–should be the norm of industrial practice in modern plants. Unfortunately, a recent survey, by Desborough and Miller has shown otherwise. This indicates that 97% of regulatory controllers are of the proportional–integral–derivative (PID) type and only 32% of the loops show "excellent" or "good" performance. Six years have passed since the first edition was published, and the practice of industrial process control is very much the same: PID controllers are widely used but poorly tuned.

This book is aimed at engineers and researchers who are looking for ways to improve controller performance. It provides a simple and yet effective method of tuning PID controllers automatically. Practical tools needed to handle various process conditions, *e.g.* load disturbance, nonlinearity and noise, are also given.

The mathematics of the subject is kept to a minimum level and emphasis is placed on experimental designs that give relevant process information for the intended tuning rules. Numerous worked examples and case studies are used to illustrate the autotuning procedure and closed-loop performance.

This book is an independent learning tool that has been designed to educate people in technologies associated with controller tuning. Most aspects of autotuning are covered, and you are encouraged to try them out on industrial control practice.

The book is divided into 12 chapters. In Chapter 1, perspectives on process control and the need for automatic tuning of PID controllers are given. The PID controller is introduced in Chapter 2. Corresponding P, I, and D actions are explained and typical tuning rules are tabulated. Chapter 3 shows how and why the relay feedback tests can be used as a means of autotuning, and an autotuning procedure is also given. A simple and an improved algorithm are explored and analytical expressions for relay feedback responses are also derived. The shape of relay feedback is discussed in Chapter 4. This gives useful information on possible model structure and ranges of model parameters. Once model structure is available, an appropriate tuning rule can be applied for improved control performance. In Chapter 5, a ramp type of relay is proposed to provide better accuracy in identifying process parameters. The improved experimental design is shown to work well for both single-input–single-output (SISO) and multivariable systems. Chapter 6 is devoted to a more common situation: multivariable systems. Experiments are devised and procedures are given for the automatic tuning of multiloop SISO controllers. Chapter 7 is devoted to a practical problem: autotuning under load disturbance. A procedure is presented to find controller parameters under load changes. The multiple-model approach is known to be effective in handling processes that are nonlinear, and Chapter 8 extends the relay feedback autotuning in a multiple-model framework. In Chapter 9, the controller monitoring problem is addressed. Again, the shape of relay feedback response gives a useful indication on the appropriateness of the tuning constant. Moreover, monitoring and retuning are completed in a single-relay feedback test. The issue of an imperfect actuator is dealt with in Chapter 10. For control valve with hyteresis, an autotuning procedure is proposed to overcome the frequently encountered problem in practice. In Chapter 11, the importance of control structure design is illustrated using a plantwide control example. Procedures for the design of the control structure and the tuning of the entire plant are given and the results clearly indicate that the combination of better process understanding and improved tuning makes the recycle plant much easier to operate. Chapter 12 summarizes the guidelines for autotuning procedures and describes when and what type of relay feedback test should be employed.

The book is based on work my students and I have been engaged in for almost 20 years to improve PID controller performance. I wrote the book because I believe strongly in the benefits of improved control, and a well-tuned PID controller is a fundamental step for improved process operation.

Acknowledgements
Thanks are due to K. J. Åström, T. Hägglund, W. L. Luyben, Q. G. Wang, I. B. Lee, and my colleague H. P. Huang, who have contributed to the development of relay feedback autotuning. Undergraduate and graduate students and postdoc fellows of NTU and NTUST have contributed to this book by their questions and interest in the subject. In particular, the continuous feedback from Walters Shen, K. L. Wu, D. M. Chang, Y. C. Cheng, Y. H. Chen, T. Thyagarajan, and R. C. Panda needs to be acknowledged. The superb editing work of Brenda Tsai and Vincent Chang made this book possible. Finally, without the understanding and support of my family, this book would not have been undertaken, or completed.

Contents

1 Introduction 1
  1.1 Scope of Process Control 1
  1.2 Proportional–Integral–Derivative Control Performance 2
  1.3 Relay Feedback Identification 5
  1.4 Conclusion 6
  1.5 References 7

2 Features of Proportional–Integral–Derivative Control 9
  2.1 Proportional–Integral–Derivative Controller 9
    2.1.1 Proportional Control 9
    2.1.2 Proportional–Integral Control 10
    2.1.3 Proportional–Integral–Derivative Control 12
  2.2 Proportional–Integral–Derivative Implementation 13
    2.2.1 Reset Windup 13
    2.2.2 Arrangement of Derivative Action 15
  2.3 Proportional–Integral–Derivative Tuning Rules 17
    2.3.1 Ziegler–Nichols Types of Tuning Rules 17
    2.3.2 Model-based Tuning 19
  2.4 Conclusion 20
  2.5 References 20

3 Relay Feedback 23
  3.1 Experimental Design 24
  3.2 Approximate Transfer Functions: Frequency-domain Modeling .. 26
    3.2.1 Simple Approach 27
    3.2.2 Improved Algorithm 30
    3.2.3 Parameter Estimation 32
    3.2.4 Examples 32
  3.3 Approximate Transfer Functions: Time-domain Modeling 36
    3.3.1 Derivation for a Second-order Overdamped System 39
    3.3.2 Results 41

x   Contents

      3.3.3   Validation 44
  3.4   Conclusion 44
  3.5   References 46

4   Shape of Relay 47
  4.1   Shapes of Relay Response 47
      4.1.1   Shapes 48
      4.1.2   Model Structures 50
          4.1.2.1 First-order Plus Dead Time 50
          4.1.2.2 Second-order Plus Small Dead Time 51
          4.1.2.3 High Order 52
  4.2   Identification 52
      4.2.1   Identification of Category 1: First-order Plus Dead Time 52
          4.2.1.1 Category 1a: True First-order Plus Dead Time 52
          4.2.1.2 Category 1b: Approximated First-order Plus Dead Time 55
      4.2.2   Identification of Category 2: Second-order Plus Small Dead Time .. 56
      4.2.3   Identification of Category 3: High order 58
      4.2.4   Validation 59
  4.3   Implications for Control 62
      4.3.1   Proportional–Integral–Derivative Control 62
          4.3.1.1 Category 1: First-order Plus Dead Time 62
          4.3.1.2 Category 2: Second-order Plus Small Dead Time 64
          4.3.1.3 Category 3: High Order 64
      4.3.2   Results 64
      4.3.3   Extension 70
          4.3.3.1 Dead-time-Dominant Process 70
          4.3.3.2 Higher Order Process 71
  4.4   Conclusion 72
  4.5   References 73

5   Improved Relay Feedback 75
  5.1   Analysis 76
      5.1.1   Ideal (On–Off) Relay Feedback 76
      5.1.2   Saturation Relay Feedback 78
      5.1.3   Potential Problem 83
  5.2   Improved Experimental Design 84
      5.2.1   Selection of the Slope of Saturation Relay 84
      5.2.2   Procedure 89
  5.3   Applications 89
  5.4   Conclusion 95
  5.5   References 96

6   Multivariable Systems 97
  6.1   Concept 97
      6.1.1   Single-input–Single-output Autotuning 97
      6.1.2   Multiple-input–Multiple-output Autotuning 99
  6.2   Theory 101

	6.2.1 Sequential Design 101
	6.2.2 Process Characteristics.. 104
	6.2.3 Sequential Identification... 108
6.3	Controller Tuning.. 111
	6.3.1 Potential Problem in Ziegler–Nichols Tuning 111
	6.3.2 Modified Ziegler–Nichols Method.. 111
	6.3.3 Performance Evaluation: Linear Model 115
6.4	Properties.. 117
	6.4.1 Convergence 117
	6.4.2 Tuning Sequence... 119
	6.4.3 Problem of Variable Pairing 120
	6.4.4 Summary of Procedure... 122
6.5	Applications.. 123
	6.5.1 Moderate-purity Column 123
	6.5.2 High-purity Column.. 124
	6.5.3 T4 Column.. 128
6.6	Conclusion 130
6.7	References 130
Appendix.. 132

7 Load Disturbance 135
7.1 Problems.. 135
7.1.1 Step Change versus Continuous Cycling............................... 135
7.1.2 Effect of Load Change on Relay Feedback Test...................... 138
7.2 Analyses 139
7.2.1 Causes of Errors... 139
7.2.2 Output-biased Relay Feedback System 142
7.2.3 Derivation of Bias Value δ_o 144
7.2.3.1 Effect of Load Disturbance 144
7.2.3.2 Opposite Effect from Output-biased Relay 146
7.3 Summary of Procedure... 148
7.4 Applications... 149
7.4.1 Linear System 150
7.4.2 Binary Distillation Column 152
7.5 Conclusion 153
7.6 References 154

8 Multiple Models for Process Nonlinearity 155
8.1 Autotuning and Local Model... 156
8.2 Model Scheduling 157
8.2.1 Takagi–Sugeno Fuzzy Model 157
8.2.1.1 Single Input Systems 158
8.2.1.2 Multiple Inputs Systems... 160
8.2.2 Selection of Scheduled Variable 162
8.3 Nonlinear Control Applications 163
8.3.1 Transfer Function System... 163
8.3.2 Tennessee Eastman Process... 167

	8.4 Conclusion	173
	8.5 References	173

9 Control Performance Monitoring 175
    9.1 Shape Factor for Monitoring..176
        9.1.1 Shapes of the Relay Feedback 176
    9.2 Performance Monitoring and Assessment 179
        9.2.1 Optimal Performance 179
        9.2.2 Proposed Monitoring and Assessment Procedure 180
            9.2.2.1 Case 1: $\tau_I/\tau = 1$ 180
            9.2.2.2 Case 2: $\tau_I/\tau > 1$ 180
            9.2.2.3 Case 3: $\tau_I/\tau < 1$ 181
        9.2.3 Illustrative Examples...184
    9.3 Applications...188
        9.3.1 Second-order Plus Dead Time Processes..................................188
        9.3.2 High-order Processes 193
    9.4 Conclusion..196
    9.5 References 196

10 Imperfect Actuators 197
    10.1 Potential Problems 197
    10.2 Identification Procedure 202
        10.2.1 Two-step Procedure 202
        10.2.2 Simultaneous Identification 205
    10.3 Applications...206
        10.3.1 Linear Systems..206
            10.3.1.1 Noise-free System..206
            10.3.1.2 Systems with Measurement Noise 207
            10.3.1.3 Load Disturbance...210
        10.3.2 Nonlinear Process 212
            10.3.2.1 Two-step Procedure 212
            10.3.2.2 Simultaneous Procedure...212
    10.4 Conclusion...216
    10.5 References 217

11 Autotuning for Plantwide Control Systems 219
    11.1 Recycle Plant 219
    11.2 Control Structure Design 222
        11.2.1 Unbalanced Schemes...222
            11.2.1.1 Column Overwork 222
            11.2.1.2 Reactor Overwork..226
        11.2.2 Balanced Scheme 227
        11.2.3 Controllability..228
        11.2.4 Operability...231
    11.3 Controller Tuning for Entire Plant 232
        11.3.1 Tuning Steps..233

        11.3.1.1 Inventory Control 233
        11.3.1.2 Ratio Control 233
        11.3.1.3 Quality Loop 235
    11.3.2 Closed-loop Performance 238
  11.4 Conclusion 242
  11.5 References 242

12 Guidelines for Autotune Procedure 245
  12.1 Process Characteristics 245
    12.1.1 The Shape 245
    12.1.2 Load Disturbance 246
    12.1.3 Nonlinearity 246
    12.1.4 Noise 247
    12.1.5 Imperfect Actuator 248
  12.2 Available Relays 248
  12.3 Specifications 249
    12.3.1 Direct Tuning 249
    12.3.2 Model-based Tuning 251
    12.3.3 Multiloop System 252
  12.4 Discussion 256
  12.5 Conclusion 257
  12.6 References 258

Index 259

1

Introduction

1.1 Scope of Process Control

Over past 50 years, "process control" has developed into a vital part of the engineering curriculum. Textbooks ranges from 600 to 1200 pages [1–3] and cover various aspects of industrial process control. It is hopeless to discuss all subjects of process control in this book. However, a brief description of the scope of process control will be given and the specific role of this book will become clear.

For continuous manufacturing, on-demand production with on-aim quality is the goal of process operation. Many factors contribute to non-smooth process operation, and controller tuning is just one of them, as shown in Figure 1.1 [4]. Starting from the most fundamental level, process variations may come from the *infrastructure* of a control system in which the signal transmission, control panel arrangement, distributed control system (DCS) selection, and DCS configuration may be the source of the problem. If the infrastructure is not the source of variation, then one may go up to the *instrumentation* level, which includes the control valve sizing, sensor selection, and transmitter span determination. It is clear that a wrongly sized control valve or an incorrectly determined transmitter span cannot provide adequate resolution in the manipulated variable or the controlled variable. It then comes to the *controller tuning* level in which inadequate controller settings may lead to oscillation in process variables, and improved controller settings is the focus of this book. If a controller retuning still cannot fix the problem, then we go to the *controller structure* level, in which one can try different types of controller. The actions in this level include: remove or add the derivative action, take out or add the integral action, use the gain scheduling, and add the dead time compensation. For example, the use of a proportional (P) only controller is often recommended for maximum flow smoothing in level control, and avoid using the derivative (D) action when the measurement is corrupted with noise. If the process variation is still significant, then it may be a problem in the *control configuration*. Experienced designers always establish loop pairings by maximizing the steady state gain between the controlled and the manipulated variables and by shortening

2   Autotuning of PID Controllers

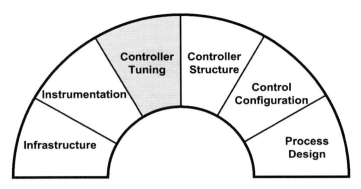

Figure 1.1. Spectrum of process operation

the response time (time constant) and dead time. Certainly, the inherent integral controllability should be maintained and the relative gain should be checked when dealing with multivariable systems. The other option is to explore the possibility of using a multivariable controller. However, one should be sure that we have enough engineering manpower for the maintenance of the much more complicated control system. Once all other possibilities are exhausted, we come to the rightmost part of the spectrum: *process design* can also be a possible cause of non-smooth operation. It has long been recognized that a process that has been design-based on some steady state economic objective will not necessarily provide good dynamic performance. This is especially true when new plants are typically designed using complex flowsheets with many streams for material recycles and for energy exchanges. The highly integrated plants generally lead to complex dynamics and difficulty in control and operation. Thus, in some cases, process redesign is required to ensure an operable process. The necessity of simultaneous design and control is advocated by Luyben and as can be seen in two recent books [4,5] and chapters of textbooks [2,3] devoted to this area. After studying the spectrum of process control, it should become clear that "controller tuning" only constitutes a fraction of the entire spectrum and it is even clearer that an improved controller tuning *cannot* solve all the problems associated with non-smooth process operation.

1.2 Proportional–Integral–Derivative Control Performance

Despite rapid evolution in control hardware, the proportional–integral–derivative (PID) controller remains the workhorse in process industries. The P action (mode) adjusts controller output according to the size of the error. The I action (mode) can eliminate the steady state offset and the future trend is anticipated via the D action (mode). These useful functions are sufficient for a large number of process applications and the transparency of the features leads to wide acceptance by the users. On the other hand, it can be shown that the internal model control (IMC) framework leads to PID controllers for virtually all models common in industrial practice [6].

Note that this includes systems with inverse responses and integrating (unstable) processes.

PID controllers have survived many changes in technology. It begins with pneumatic control, through direct digital control to the DCS. Nowadays, the PID controller is far different from that of 50 years ago. Typically, logic, function block, selector and sequence are combined with the PID controller. Many sophisticated regulatory control strategies, override control, start-up and shut-down strategies can be designed around the classical PID control. This provides the basic means for good regulatory, smooth transient, safe operation and fast start-up and shut-down. Moreover, even with model predictive control (MPC), PID controllers still serve as the fundamental building block at the regulatory level. The computing power of microprocessors provides additional features, such as automatic tuning, gain scheduling and model switching, to the PID controller. Eventually, all PID controllers will have the above-mentioned intelligent features.

In process industries, more than 97% of the regulatory controllers are of the PID type [7]. Most loops are actually under PI control (as a result of the large number of flow loops). More than 60 years after the publication of the Ziegler–Nichols tuning rule [8] and with the numerous papers published on the tuning methods since, one might think that the use of PID controllers has already met our expectations. Unfortunately, this is not the case. Surveys of Bialkowski [9], Ender [10], McMillan [11], Hersh and Johnson [12], and Desborough and Miller [7] show that:

1. Pulp and paper industry over 2000 loops [9]
   - Only 20% of loops worked well (*i.e.* less variability in the automatic mode over the manual mode).
   - 30% gave poor performance due to poor controller tuning.
   - 30% gave poor performance due to control valve problems (*e.g.* control valve stick-slip, dead band, backlash).
   - 20% gave poor performance due to process and/or control system design problems.

2. Process industries [10]
   - 30% of loops operated on manual mode.
   - 20% of controllers used factory tuning.
   - 30% gave poor performance due to sensor and control valve problems.

3. Chemical process industry [11]
   - Half of the control valves needed to be fixed (results of the Fisher diagnostic valve package).
   - Most poor tuning was due to control valve problems.

4. Manufacturing and process industries [12]
   - Engineers and managers cited PID controller tuning as a difficult problem.

5. Refining, chemicals, and pulp and paper industries over 26,000 controllers [7]

- Only 32% of loops were classified as "excellent" or "acceptable".
- 32% of controllers were classified as "fair" or "poor", which indicates unacceptably sluggish or oscillatory responses.
- 36% of controllers were on open- loop, which implies that the controllers were either in manual or virtually saturated.
- PID algorithms are used in vast majority of applications (97%). For the rare cases of complex dynamics or significant dead time, other algorithms are used. MPC acts less as a multivariable regulatory controller and more like a dynamic optimizer.

Surveys indicate that the process control performance is, indeed, "not as good as you think" [10], and the situation remains pretty much the same a decade later [7]. The reality leads us to reconsider the priorities in process control research. First, an improved process and control configuration redesign (*e.g.* selection and pairing of input and output variables) can improve control performance. As mentioned earlier, simultaneous design and control should be taken seriously to alleviate the problem of a small operating window and the requirement for sophisticated control configuration. Second, control valves contribute significantly to the poor control performance. It is difficult, if not impossible, to replace or to restore all the control valves to the expected performance. In other words, in many cases, this is a fact we have to face (*e.g.* dead band, stick-slip, *etc.* [13]). One thing we can do is to devise a diagnostic tool to identify potential problems in control valves. We have seen the beginning of research effort in this direction [14–17]. Third, and probably the easiest way to improve control performance, is to find appropriate tuning constants for PID controllers.

Sixty years after Ziegler and Nichols published their famous tuning rule, numerous tuning methods have been proposed in the literature. We do expect that engineers have gained proficiency in the design of simple PID controllers. The reality indicates that this is simply not the case. Moreover, the structure of current leaner corporations does not offer much opportunity to improve the situation. Another factor is the time required for the tuning of many *slow* loops (*e.g.* temperature loops in high-purity distillation columns). On many occasions, engineers simply do not have the luxury and patience to tune a loop over a long period of time (not being able to complete the task in a shift). It then becomes obvious that the PID controller with an automatic tuning feature is an attractive alternative for better control. That is, instead of continuous adaptation, the controller should be able to find the tuning parameters by itself: it is an autotuner.

Table 1.1 shows the current trend where major vendors provide one type or more autotuners in their products [18]. Identification methods include: open- or closed- loop step tests (step), relay feedback test (relay), and possibly pseudo-random binary signal (PRBS). The feature of gain-scheduling is also available in many of the products.

In devising such an automatic tuning feature, several factors should be considered:

Introduction 5

Table 1.1. Autotuners from different vendors

Manufacturer	Identification method	Gain scheduling
ABB	Step/relay	Yes
Emerson Process Management	Relay	Yes
Foxboro	Step	No
Honeywell	Step	Yes
Siemens	Step	Yes
Yokogawa	Step	Yes

1. Control tuning can improve the performance, but it should be recognized that good tuning can only solve *part* of the problem.
2. The experimental design for system identification becomes rather important, since we are not able to keep all the control valves in perfect condition.
3. The system identification step should be time efficient. This is rather useful for many slow industrial processes.

1.3 Relay Feedback Identification

System identification plays an integral part in automatic tuning of the PID controller. Based on the information obtained, the methods for identification can be classified into the frequency-domain and time-domain approaches.

The time-domain approaches generate responses from step or pulse tests [2,3]. The characteristics of the process response are then utilized to back-calculate the parameters of an assumed process model [19]. The step tests can be performed in open-loop (manual mode) or closed-loop mode (while controller is working). The open-loop step test is fairly straightforward. However, it is vulnerable to load disturbances, especially for systems with large time constants. Moreover, the behavior of the control valve is not fully tested in the experiment. The closed-loop step tests, on the other hand, can shorten the time for experiment. But we have to choose a set of controller parameters in order to generate oscillatory (underdamped) responses [19]. The process model is then approximated from the damping behavior. The pattern recognition controller [20,21] is a typical example. Since step-like change is involved, it is not expected to work well for highly non-linear systems, (*e.g.* high purity distillation columns [22]).

Another category is the Ziegler–Nichols type of experimental design. Probably the more successful part of the Ziegler–Nichols method is *not* the tuning rule itself. Rather, it is the identification procedure: a way to find the important process information, ultimate gain K_u and ultimate frequency ω_u. This is often referred to as the trial-and-error procedure [2,3]. A typical approach can be summarized as follows:

1. Set the controller gain K_c at a low value, perhaps 0.2.
2. Put the controller in the automatic mode.
3. Make a small change in the set point or load variable and observe the response. If the gain is low, then the response will be sluggish.
4. Increase the gain by a factor of two and make another set point or load change.
5. Repeat step 4 until the loop becomes oscillatory and continuous cycling is observed. The gain at which this occurs is the ultimate gain K_u, and the period of oscillation is the ultimate period P_u ($P_u = 2\tau / \omega_u$).

This is a simple and reliable approach to obtain K_u and ω_u. The disadvantage is also obvious: it is time consuming. The present-day version is the relay feedback test proposed by Åström and Hägglund [23]. First, a continuous cycling of the controlled variable is generated from a relay feedback experiment and the important process information, K_u and ω_u, can be extracted directly from the experiment. The information obtained from the relay feedback experiment is exactly the same as that from the conventional continuous cycling method. It should be noticed that the relay feedback is an old and useful technique for feedback control, as can be seen from earlier results [24,25], and, here, a new meaning is assigned to the relay feedback. However, an important difference is that the sustained oscillation is generated in a *controlled* manner (*e.g.* the magnitude of oscillation can be controlled) in the relay feedback test. Moreover, in virtually all cases, this is a very efficient way, *i.e.* a one-shot solution, to generate a sustain oscillation. Applications of the Åström and Hägglund autotuner are found throughout process industries using single-station controllers or a DCS (Table 1.1). The success of this autotuner is due to the fact that the identification and tuning mechanism are so *simple* that operators understand how it works. It also works well in slow and highly nonlinear processes [22]. Over the past two decades, extensive research has been done on relay feedback tests. Refinements on the accuracy and improvements on the experimental design have been made. Discussions about potential problems, extensions to multivariable systems and incorporation of gain scheduling have also been reported. Luyben brings the autotuner to another level in which the "shape" of relay feedback can be utilized to identify the model structure. This motivates us for the revision. It is our view that the relay-feedback-based autotuners now can provide the necessary tools to improve control performance in a reliable way.

1.4 Conclusion

In this chapter we clearly define the scope of process control, and one should realize that the controller tuning only constitutes a fraction of the process operation problems. Surveys indicate that the PID controller is the major controller in process industries. After many years of experience, the control loops, often thought too simple, do not perform as well as one might expect. The failure comes from the lack of the required knowledge to maintain the control loops, to tune the controllers, to design an appropriate process for control and to design a suitable control configuration for a given process. Poor control performance may have many different causes. However, obtaining good tuning is always the most cost-effective way to improve control. You should recognize that controllers are working with imperfect valves, noisy sensors and frequent load disturbances. These factors have to be taken into account when you are designing the experiment to find controller parameters.

1.5 References

1. Ogunnaike BA, Ray WH. Process dynamics, modeling, and control. New York: Oxford University Press; 1994.
2. Luyben WL, Luyben ML. Essentials of process control. New York: McGraw-Hill; 1997.
3. Seborg DE, Edgar TF, Mellichamp DA. Process dynamics and control; 2nd ed. New York: John Wiley & Sons; 2004.
4. Luyben WL, Tyreus BD, Luyben ML. Plantwide control. New York: McGraw-Hill; 1999.
5. Seferlis P, Georgiadis MC. The integration of process design and control. 2nd ed. Amsterdam: Elsevier; 2004.
6. Morari M, Zafiriou E. Robust process control. Prentice Hall: Englewood Cliff; 1989.
7. Desborough L, Miller R. Increasing customer value of industrial control performance monitoring-Honeywell experience. In: 6th Int. Conf. Chemical Process Control, AIChE Symp. Series 326; Rawlings JB, Ogunnaike BA, Eaton JW. Eds. AIChE: New York; 2002.
8. Ziegler JG, Nichols NB. Optimum settings for automatic controllers. Trans. ASME 1942;12:759.
9. Bialkowski WL. Dream vs Reality: A view from both sides of the gap. Pulp Paper Can 1993;94:19.
10. Ender DB. Process control performance: Not as good as you think. Control Eng 1993;40:180.
11. McMillan GK. Tuning and control loop performance. Instrument Society of America: Research Triangle Park; 1994.
12. Hersh MA, Johnson MA. A study of advanced control systems in the workplace. Control Eng. Prac. 1997;5:771.
13. Cheng YC, Yu CC. Relay feedback identification for actuators with hysteresis. Ind. Eng. Chem. Res. 2000;39:4239.

14. Huang B, Shah SL. Control loop performance assessment: Theory and applications. Springer-Verlag; 1999.

15. Choudhury MAAS, Shah SL, Thornhill NF. A data-driven model for valve stiction. ADCHEM 2003; 2004 Jan. 11–14; Hong Kong; 2004.

16. Manabu K, Maruta H, Kugemoto H, Shimizu K. Practical model and detection algorithm for valve stiction. The 7th Int. Conference on Dynamics and Control of Process Systems (DYCOPS-7); 2004 July 5–7; Boston; 2004.

17. Rossi M, Scali C. A comparison of techniques for automatic detection of stiction: simulation and application to industrial data. J. Proc. Cont. 2005;15:505.

18. Hang CC, Lee TH, Ho TH. Adaptive control. Instrument Society of America: Research Triangle Park; 1993.

19. Yuwana M, Seborg DE. A new method for on-line controller tuning. AIChE J. 1982;28:434.

20. Bristol EH. Pattern recognition: An alternative to parameter adaptive PID controller. Automatica 1977;13:197.

21. Cao R, McAvoy TJ. Evaluation of pattern recognition adaptive PID controller. Automatica 1990;26:797.

22. Luyben WL. Derivation of transfer functions for highly nonlinear distillation columns. Ind. Eng. Chem. Res. 1987;26:2490.

23. Åström KJ, Hägglund T. Automatic tuning of simple regulators with specifications on phase and amplitude margins. Automatica 1984;20:645.

24. Tsypkin IZ. Relay control systems. Cambridge: Cambridge University Press; 1984.

25. Atherton DP. Nonlinear control engineering. New York: Van Nostrand Reinhold; 1982.

26. Luyben WL. Getting more information from relay feedback tests. Ind. Eng. Chem. Res. 2001;40:4391.

2

Features of Proportional–Integral–Derivative Control

2.1 Proportional–Integral–Derivative Controller

The proportional–integral–derivative controller consists of three simple actions, *i.e.* P, I, and D actions. Let us use a heat exchanger (a cooler to be exact) example to illustrate these three functions. Figure 2.1 shows the inlet stream is cooled to a specific temperature by exchanging heat with cooling water. So the controlled variable is the heat exchanger outlet temperature and the manipulated variable is the cooling water flow rate. The heat exchanger outlet temperature is measured using a thermocouple, and then it is converted into a signal, generally called the process variable (PV), which is compatible with the control system (typically in the range of 4–20 mA). The PV is compared with the set point (SP) and the controller output (CO) is generated based on the control algorithm. The controller output is further converted to an air pressure signal to drive the valve. In doing so, the real cooling water flow rate is set according to the stem position (determined by CO), size of the valve, pressure drop across the valve, and the valve characteristic. The feedback controller generates its move based on the error E between the SP and PVs, $E(t)=SP(t)-PV(t)$.

2.1.1 Proportional Control

The P controller changes its output CO in direct proportion to the error signal E.

$$CO = Bias + K_c(SP - PV) \qquad (2.1)$$

The bias signal is the value of the controller output when there is no error. This is an intuitive and simple action which is quite similar to human behavior. Whenever we are far away from our goal, we make a larger adjustment, and when we come close to the target, a smaller step is taken. Here, K_c is called the *controller gain*, an adjustable parameter. Figure 2.2 shows the responses of a P controller with three values of K_c for a step decrease in the heat exchanger inlet temperature. It becomes

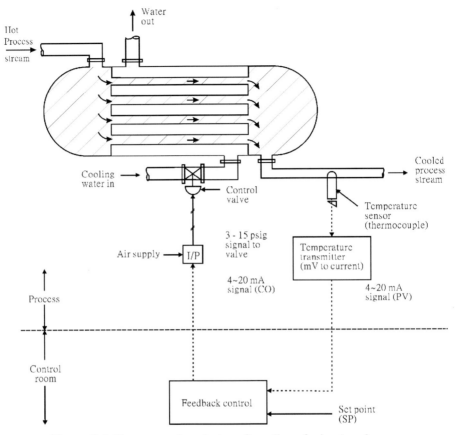

Figure 2.1. Process and control configuration of a heat exchanger

obvious that steady state errors (offset) exist for the P control. The responses indicate that an increase in the controller gain K_c can reduce the offset, but the response tends to be oscillatory. Certainly, when K_c is set to zero, the process is effectively open loop. To summarize the behavior of P control, we have: (1) it is a simple and intuitive, and (2) a steady state offset exists.

2.1.2 Proportional–Integral Control

In order to eliminate steady state offset, the I action is often included. I action moves the control valve in direct proportion to the time integral of the error. The resultant PI controller can be expressed as

$$CO = Bias + K_c \left((SP - PV) + \frac{1}{\tau_I} \int (SP - PV) dt \right) \qquad (2.2)$$

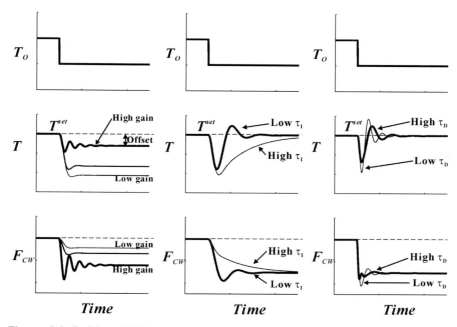

Figure 2.2. P, PI, and PID control performance using different controller settings for a step decrease in the heat exchanger inlet temperature T_o

Here, we have a second tuning parameter τ_I, which is called the *reset time* or the *integral time* with units of time (typically minutes). The PI controller equation indicates that the CO will keep changing until the difference between the SP and PV diminishes, *i.e.* E=0. This can be viewed as a relentless effort to meet the target by changing the input effort. In other words, the CO will not *rest* until the steady state error becomes zero. The integral action usually degrades the closed-loop performance. In a control notation, it introduces a 90° phase lag into the feedback loop. But the integral action is often needed for its ability to eliminate steady state offset. Figure 2.2 shows that, with I action, the heat exchanger outlet temperature does return to the set point. A smaller τ_I speeds up the temperature response while becoming a little oscillatory. It should be noticed that most of controllers (~70%) in industry are PI controllers. Instead of using the controller algorithm explicitly, most of the controller manuals express the PI controller in terms of a Laplace transformation (this is probably one of the few Laplace transformations you need to recognize when working in industry).

$$PI = \frac{CO}{E} = K_c \left(1 + \frac{1}{\tau_I s}\right) = K_c \left(\frac{\tau_I s + 1}{\tau_I s}\right) \qquad (2.3)$$

2.1.3 Proportional–Integral–Derivative Control

The D action uses the *trend* of the process variable to make necessary adjustments. The process trend is estimated using the derivative of the error signal with respect to time. The *ideal* PID has the following form:

$$CO = Bias + K_c \left((SP-PV) + \frac{1}{\tau_I} \int (SP-PV)dt + \tau_D \frac{d(SP-PV)}{dt} \right) \quad (2.4)$$

Here, the third tuning parameter τ_D is the *derivative time* with units of time. It may be intuitive, appealing that the "process trend" can be incorporated into a control algorithm. We use these types of trend (or derivative) in numerical methods, *e.g.* Newton–Raphson method, and in stocks selling and buying. In theory, adding derivative action should always improve the dynamic response, and it should be the preference over the PI controller. The Laplace transformation of the *ideal* PID controller can be expressed as

$$PID_{ideal} = K_c \left(1 + \frac{1}{\tau_I s} + \tau_D s \right) \quad (2.5)$$

However, the ideal PID control algorithm has rarely been implemented in practice. Instead, a *filtered* D action is often used. The following is the *parallel* form of PID control with filtered D action:

$$PID_{parallel} = K_c \left(1 + \frac{1}{\tau_I s} + \frac{\tau_D s}{\alpha \tau_D s + 1} \right) \quad (2.6)$$

where α typically takes a value of 1/10. Figure 2.2 clearly shows that the PID controller outperforms the PI controller in the noise-free condition. But too large a τ_D will lead to significant oscillation in the controlled variable. However, when the process measurement is corrupted with noise, we have a completely different behavior, especially in the manipulated variable. Figure 2.3 indicates that the control valve is banging up and down, when we have fluctuating process measurements. This is certainly not desirable from the maintenance perspective. This also confirms why most controllers in industry are PI controllers, instead of PID controllers. This is typically true in chemical process industries when many flow loops are installed.

The P–I–D actions can be summarized as follows. P action is intuitive and effective, I action is relentless and offset free, and D action is the trend finder, but noise sensitive. After understanding the characteristic of each action, one should find the right combination of P–I–D actions for the controller to achieve good control performance.

Features of PID Control   13

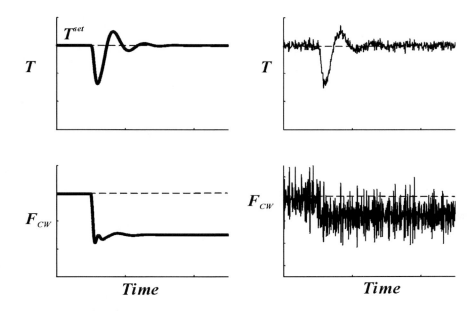

Figure 2.3. Load responses of PID controller with noise-free and noisy measurements

2.2 Proportional–Integral–Derivative Implementation

Two implementation issues for PID control are addressed. One is the anti-reset-windup associated with controllers with I action, and the other is the D action arrangement in a PID controller.

2.2.1 Reset Windup

The reset windup is an important and realistic problem in process control. It may occur whenever a controller contains the I action. When a sustained error occurs, the I term becomes quite large and the CO eventually goes beyond saturation limits (CO greater than 100% or less than 0%). Because all actuators have limitations, *e.g.* the flow through a control valve is limited by its size, and if the controller is asking for more than the actuator can deliver, there will be a difference between the CO and the actual control action (CO_A). When this happens, the controller is effectively disabled, because the valve remains unchanged, *e.g.* in full-open position. Not recognizing this circumstance, the controller continues to perform numerical integration, and the CO becomes even larger. It then requires (1) the error changing sign and (2) a long time to digest all the accumulated integrand, before the control valve moves away from the saturation limits. This is known as the *reset windup*. The consequence is a long transient and large overshoot in the controlled variable [1,2]. The reset windup may occur as a consequence of large disturbances or it may be caused by large SP changes, *e.g.* during the start-up of a batch process. Windup may also arise when the override control is used and so we have two controllers with only one control valve.

Conceptually, reset windup can be prevented by turning off the I action whenever the CO saturates. Many antiwindup methods have been proposed for different types of controller and for single-variable and multivariable systems [3,4]. One simple and effective approach for the integral windup is shown in Figure 2.4. The scheme involves a negative feedback loop around the I action with the CO in the loop. At normal operation (without saturation), the CO is equal to the actual control action CO_A, *i.e.* $CO = CO_A$, the feedback path disappears and the I action is in place. The actuator model is simply

$$CO_A = \begin{cases} 0 & CO<0 \\ CO & 0 \le CO \le 1 \\ 1 & CO>1 \end{cases} \qquad (2.7)$$

When the I action winds up, the actual control action remains unchanged, *e.g.* $CO_A=1$, and it can be treated as a reference value which is different from the controller output. Thus, the antiwindup scheme is best described by the following Laplace transformed relationship according to Figure 2.4:

$$CO(s) = \frac{1}{\tau s + 1} CO_A(s) + \frac{\tau}{\tau s + 1} E_I(s) \qquad (2.8)$$

It becomes clear that the feedback loop tends to drive the CO to the actual control action following a first-order dynamics. The adjustable parameter τ is called the tracking time constant, and, typically, it is set to a small value. The antiwindup scheme now becomes a standard feature in commercial PID controllers.

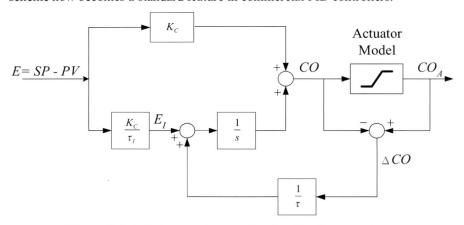

Figure 2.4. Antiwindup scheme with a tracking time constant τ

2.2.2 Arrangement of Derivative Action

For PI controllers, the proportional and the integral actions are additive, and the PI algorithm is universally used in all controllers. Unlike the PI controller, the PID controller appears to have many different forms. The two most common types are shown in Figure 2.5. The first type of PID controller has the three actions working additively. The continuous transfer function is given in Equation 2.6. It is called descriptively as the "parallel" form of PID controller. We label this as $PID_{parallel}$. The second type of PID controller can be expressed in terms of the following transfer function:

$$PID_{series} = K_c \left(\frac{\tau_I s + 1}{\tau_I s} \right) \left(\frac{\tau_D s}{\alpha \tau_D s + 1} \right) \qquad (2.9)$$

This type of PID controller is known as the "series" form of PID controller, as can be seen from the equation and the block diagram arrangement in Figure 2.5. It was used in early analog controllers and has been implemented digitally in modern DCSs. Some of the popular tuning methods, *e.g.* Ziegler–Nichols [5], Tyreus and Luyben [6], and Luyben [7], are based on this algorithm. They also assume that the derivative filter parameter had a value of $\alpha=0.1$. And yet another type of PID controller is the four-parameter PID controller, which is derived from the internal model control [8]. This is denoted as the "IMC" form of PID controller, PID_{IMC}. The following is the transfer function of the IMC PID controller:

(A) Parallel

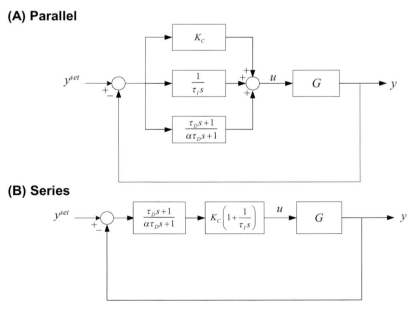

(B) Series

Figure 2.5. Parallel and series types of PID controller

$$PID_{IMC} = K_c \left(1 + \frac{1}{\tau_I s} + \tau_D s\right) \frac{1}{\tau_F s + 1} \quad (2.10)$$

Unlike the previous two types of PID controller, where α is a fixed value, the fourth tuning constant τ_F is also an adjustable parameter. These different PID forms clearly indicate that the settings of a PID controller depend on the algorithm used. The settings for the "series" and "parallel" can be very different, and one should always be aware of which algorithm the tuning rule is based on. However, the controller parameters for one algorithm can be transformed to the other as shown in Table 2.1. For example, the tuning constants of the "parallel" form can be transformed into the settings for the "series" form PID [2] and *vice versa*. Similarly, the relationship between the settings of PID_{IMC} and $PID_{parallel}$ can also be derived.

Another commonly used PID implementation is to take the derivative on the PV, instead of the error E = SP – PV, as shown in Figure 2.6. This can be understood, because a pure derivative of a step change corresponds to an impulse, and this implies a full swing of the control valve in an extremely short period of time, which is not desirable in practice. This is also known as the derivative kick [9]. This arrangement in Figure 2.6 is a standard feature in most commercial controllers.

Along this line, the PID controller can be extended further to a five-parameter controller by addressing the effects of derivative and proportional kicks [1,9].

$$CO = Bias + K_c \left((\beta \cdot SP - PV) + \frac{1}{\tau_I} \int (SP - PV) dt + \tau_D \frac{d(\gamma \cdot SP - PV)}{dt}\right) \quad (2.11)$$

Here, the two SP weightings, β and γ, are two additional adjustable parameters ranging from 0 to 1. This is often called the beta–gamma controller. The control algorithm in Equation 2.11 allows independent SP weightings in the proportional and derivative terms. To eliminate derivative kick, γ is set to zero, and similarly, β

Table 2.1. Interchangeable controller settings for different forms of PID controllers

$PID_{parallel} \rightarrow PID_{series}$*	$PID_{series} \rightarrow PID_{parallel}$	$PID_{IMC} \rightarrow PID_{parallel}$
$K_{c,series} = \frac{K_c}{2}\left(1 + \sqrt{1 - \frac{4\tau_D}{\tau_I}}\right)$	$K_{c,parallel} = K_c \frac{\tau_I + \tau_D}{\tau_I}$	$K_{c,parallel} = K_c \frac{\tau_I - \tau_F}{\tau_I}$
$\tau_{I,series} = \frac{\tau_I}{2}\left(1 + \sqrt{1 - \frac{4\tau_D}{\tau_I}}\right)$	$\tau_{I,parallel} = \tau_I + \tau_D$	$\tau_{I,parallel} = \tau_I - \tau_F$
$\tau_{D,series} = \frac{2\tau_D}{1 + \sqrt{1 - \frac{4\tau_D}{\tau_I}}}$	$\tau_{D,parallel} = \frac{\tau_I \tau_D}{\tau_I + \tau_D}$	$\alpha = \frac{(\tau_I - \tau_F)\tau_F}{\tau_I \tau_D - (\tau_I - \tau_F)\tau_F}$
		$\tau_{D,parallel} = \frac{\tau_F}{\alpha}$

* Valid for $\tau_I/\tau_D \geq 4$

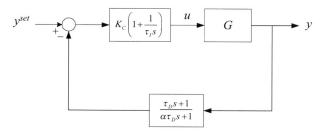

Figure 2.6. PID controller arrangement to avoid derivative kick

is set to zero to avoid proportional kick. The PID implementation in Figure 2.6 corresponds to the case of $\beta = 1$ and $\gamma = 0$. However, five tuning parameters, in general, may be prove to be too many, especially when one may be dealing with thousands of controllers in a plant.

2.3 Proportional–Integral–Derivative Tuning Rules

Since the introduction of Ziegler–Nichols tuning in 1942, PID tuning has led to remarkable research activity over the past 60 years. A recently published book, *Handbook of PI and PID Controller Tuning Rules* by O'Dwyer [10], has compiled 245 tuning rules, among which 104 are for PI and 141 for PID controllers. While been impressed by the tremendous effort, we might just run into a situation Martin [11] encountered almost 30 years ago when trying to find a useful cubic equation of state to describe vapor–liquid equilibrium. The scenario is similar to the queen in Snow White, who asked, "Mirror, mirror on the wall, who's the fairest of them all?" Now, we stand here and ask: "PID tuning rule – which?" The answer may not be the one you like to hear, but the truth is: it really depends on your processes (*e.g.* the process type, the order, the parameters, the nonlinearity, the uncertainties, *etc.*) In this section, we focus on two types of tuning rule. One is the Ziegler–Nichols type of (*i.e.* ultimate gain- and ultimate period-based) tuning rules, and the other is the model-based tuning rules. Furthermore, the process type is limited to the first-order plus dead time (FOPDT) system and its variants.

$$G(s) = \frac{K_p e^{-Ds}}{\tau s + 1} \qquad (2.12)$$

where K_p is the steady state gain, D is the dead time, and τ is the time constant.

2.3.1 Ziegler–Nichols Types of Tuning Rule

By Ziegler–Nichols type of tuning rules, we mean the PID settings are expressed explicitly in terms of the ultimate gain K_u and the ultimate period P_u from a sustained oscillation. Ziegler–Nichols tuning is still popular in control engineering practice. It works reasonably well for some loops but tends to be too underdamped for many process control applications. Many *modified* versions of Ziegler–Nichols tuning have been proposed over past 60 years [10,12]. A frequency domain interpretation of the Ziegler–Nichols method is also given [13,14].

The original Ziegler–Nichols tuning rule, expressed in terms of Ku and Pu as shown in Table 2.2, is extremely simple to use when the ultimate properties are available. One should be aware that the PID settings are for the "series" arrangement. The Ziegler–Nichols settings work well for a small range of dead time to time constant ratio D/τ, and the performance starts to degrade for $D/\tau < 0.2$ and $D/\tau > 2$ [8]. A more conservative tuning rule is proposed by Tyreus and Luyben [6,7]. The Tyreus–Luyben tuning rule is derived based on the integrator plus dead time system, $G = K_p e^{-Ds}/s$. This setting works well for the *time-constant-dominant processes*, and our experience shows that it also works well for interacting multivariable systems. At the other end of the spectrum, *dead-time-dominant processes*, the Ciancone–Marlin tuning rule is suggested [15,16]. This specific Ciancone–Marlin setting is obtained by examining pure dead time processes. The detuning factors are obtained by converting the process-parameter-based tuning constants to K_u- and P_u-related settings. The Ciancone–Marlin detuning factors are quite different from the Ziegler–Nichols settings. Here, we have a more conservative proportional action and a more aggressive integral action, as can be seen in Table 2.2. Note that the "series" PID controller is assumed for all three tuning rules.

Table 2.2 clearly shows that the selection of the "right" tuning rule is really process dependent. The D/τ ratio is a good measure to locate the appropriate one. Moreover, as will be explained in Chapter 4, the shape of the relay feedback responses gives a good indication of this D/τ ratio.

Table 2.2. Ziegler–Nichols types of tuning: (1) the original, (2) for time-constant-dominant processes and (3) for dead-time-dominant processes

Ziegler–Nichols	K_c	τ_I	τ_D*	Remarks
P	$K_u/2$	–	–	Recommended for $0.2 < D/\tau < 2$
PI	$K_u/2.2$	$P_u/1.2$	–	
PID	$K_u/1.7$	$P_u/2$	$P_u/8$	
Tyreus–Luyben	K_c	τ_I	τ_D	
PI	$K_u/3.2$	$P_u/0.45$	–	Derived from $K_p e^{-Ds}/s$
PID	$K_u/2.2$	$P_u/0.45$	$P_u/6.3$	
Ciancone–Marlin	K_c	τ_I	τ_D	
PI	$K_u/3.3$	$P_u/4$	–	Based on $K_p e^{-Ds}$
PID	$K_u/3.3$	$P_u/4.4$	$P_u/8.1$	

* "Series" form of PID

2.3.2 Model-based Tuning

A comprehensive model-based design method, IMC, is proposed by Morari and coworkers [8,17]. The IMC design consists of the following steps:

1. Assume a process model $\tilde{G}$.
2. Factor the model into an invertible part $\tilde{G}_-$ and a non-invertible part $\tilde{G}_+$. The non-invertible part contains the dead time and right-half-plane zeros.
3. Design the IMC controller as $G_c = \tilde{G}_-^{-1} \cdot F$. Here, F(s) is a low-pass filter with a gain of unity which makes the controller proper, e.g. $F = 1/(\lambda s + 1)^n$. λ is the user-specified tuning parameter.
4. Tranform the IMC controller $G_c(s)$ into a controller $K(s)$ in the conventional feedback structure (e.g. Figure 2.5), i.e. $K(s) = G_c/(1 - \tilde{G}G_c)$.

The IMC approach offers a systematic way to identify the controller structure, and, probably more importantly, it can also be used to derive tuning parameters for the PID controller [8], usually known as IMC-PID tunings. As of the inverse-based nature, it is clear that a high-order model (e.g. more than second order) leads to a high-order controller which may not be useful in practice, e.g. imaging high-order derivatives of measurement noises. Rivera *et al.* [8], Chien and Fruehauf [18], and Bequette [19] give a comprehensive list of IMC-PID tunings for different types of model. Here, the IMC-PID tunings for the first-order plus dead time, integrator plus dead time (time constant dominates), and pure dead time (dead time dominate) processes are given in Table 2.3. Both PI and PID settings are given, and recommendations for the filter time constant are also given.

Table 2.3. The IMC-PID controller settings for: (1) the first-order plus dead time, (2) integrator plus dead time, and (3) pure dead time processes

Model	K_c	τ_I	τ_D *	τ_F **	Remark
$\dfrac{K_p e^{-Ds}}{\tau s + 1}$	$\dfrac{\tau + D/2}{K_p \lambda}$	$\tau + D/2$	–		$\lambda > 1.7D$ $\lambda > 0.2\tau$
$\dfrac{K_p e^{-Ds}}{\tau s + 1}$	$\dfrac{\tau + D/2}{K_p(\lambda + D/2)}$	$\tau + D/2$	$\dfrac{\tau D}{2\tau + D}$		$\lambda > 0.8D$ $\lambda > 0.2\tau$
$\dfrac{K_p e^{-Ds}}{s}$	$\dfrac{2\lambda + D}{K_p(\lambda + D)^2}$	$2\lambda + D$	–		$\lambda > D$
$\dfrac{K_p e^{-Ds}}{s}$	$\dfrac{2}{K_p(\lambda + D/2)}$	$2\lambda + D$	$\dfrac{\lambda D + D^2/4}{2\lambda + D}$		$\lambda > D$
$K_p e^{-Ds}$	$\dfrac{D}{K_p(2\lambda + D)}$	$D/2$	–		$\lambda > D$
$K_p e^{-Ds}$	$\dfrac{D}{K_p(4\lambda + D)}$	$D/2$	$D/6$	$\dfrac{2\lambda^2 - D^2/6}{4\lambda + D}$	$\lambda > 1.7D$

* "Parallel" form of PID
** IMC PID

2.4 Conclusion

In this chapter we explore the features of the PID control. The selection of a P, PI, or PID controller really depends on the process requirement (*e.g.* tolerable for steady state offset) and noise content of the process variable. One thing is certain, that including all three actions does not necessarily imply improved performance. The implementation issues associated with PI and PID controllers are discussed. The problem of the reset windup is explained and a solution is given. The types and arrangement of PID controllers are described next. Unlike a PI controller, one should be always aware of the type of PID algorithm used in your controller and correct tuning parameters can be set. Finally, two types of PID tuning are presented. One is the familiar Ziegler–Nichols type of tuning, in which the controller settings are expressed explicitly in terms of ultimate gain and ultimate period. One should recognize that no single tuning rule works well for all systems. One should also always distinguish the process characteristic (*e.g.* time-constant- or dead-time-dominant process, or in between) and apply an appropriate tuning rule. The IMC tunings are also given for first-order type processes. This can be useful if a low-order process model becomes available.

2.5 References

1. Åström KJ, Hägglund T. PID controllers: Theory, design, and tuning. Instrument Society of America: Research Triangle Park; 1995.
2. Smith CA, Corripio AB. Principles and practice of automatic process control. 2nd ed. New York: John Wiley & Sons; 1997.
3. Hanus R, Kinnaert M, Henrotte JL. Conditioning technique, a general anti-windup and bumpless transfer method, Automatica 1987;23:729.
4. Kothare MV, Campo PJ, Morari M, Nett C. A unified framework for study of anti-windup designs. Automatica 1994;30:1869.
5. Ziegler JG, Nichols NB. Optimum settings for automatic controllers. Trans. ASME 1942;12:759.
6. Tyreus BD, Luyben WL. Tuning PI controllers for integrator/dead time processes. Ind. Eng. Chem. Res. 1992;31:2625.
7. Luyben WL. Tuning proportional–integral–derivative controllers for integrator/dead time processes. Ind. Eng. Chem. Res. 1996;35:3480.
8. Rivera DE, Morari M, Skogestad S. Internal model control 4 PID controller design. Ind. Eng. Chem. Proc. Des. Dev. 1986;25:252.
9. Seborg DE, Edgar TF, Mellichamp DA. Process dynamics and control. 2nd ed. New York: John Wiley & Sons; 2004.

10. O'Dwyer A. Handbook of PI and PID controller tuning rules. London: Imperial College Press; 2003.
11. Martin JJ. Cubic equation of state – Which?. Ind. Eng. Chem. Fundam. 1979;18:81.
12. Luyben WL, Luyben ML. Essentials of process control. New York: McGraw-Hill; 1997.
13. Åström KJ, Hägglund T. Automatic tuning of PID controllers. Instrumentation Society of America: Research Triangle Park; 1988.
14. Zhuang M, Atherton DP. Automatic tuning of optimum PID controllers. IEE Proc. D 1993;140:216.
15. Luyben WL. Effect of derivative algorithm and tuning selection on the PID control of dead time processes. Ind. Eng. Chem. Res. 2001;40:3605.
16. Marlin TE. Process control. 2nd ed. New York: McGraw-Hill; 2000.
17. Morari M, Zafirou E. Robust process control. Prentice Hall: Englewood Cliff; 1989.
18. Chien IL, Fruehauf PS. IMC tuning to improve controller performance. Chem. Eng. Prog. 1990;86(10):33.
19. Bequette BW. Process control: Modeling, design, and simulation. Prentice-Hall: Upper Saddle River; 2003.

3
Relay Feedback

Åström and Hägglund [1] suggest the relay feedback test to generate sustained oscillation as an alternative to the conventional continuous cycling technique. It is very effective in determining the ultimate gain and ultimate frequency. Luyben [2] popularizes the relay feedback method and calls this method "ATV" (autotune variation). The acronym also stands for *all-terrain vehicle*, since ATV provides a useful tool for the rough and rocky road of system identification.

As pointed out by Luyben, the motivation for using the relay feedback (ATV) has grown out of a study of an industrial distillation column. The distillation column is an important unit in chemical process industries. It is rather difficult to obtain a linear transfer function model for highly nonlinear columns. Attempts have been made using step or pulse tests. Unfortunately, the system results in an extremely long time constant, *e.g.* $\tau \approx 870$ h [2]. Moreover, very large deviations occur in the linear model as the size or direction of the input is changed. Simulation studies also reveal that, sometimes, very small changes of magnitude (less than 0.01%) have to be made to get an accurate linear model. This immediately rules out the use of this kind of input design in real plants because plant data are never known to anywhere near this order of accuracy. Luyben shows that the simple relay feedback tests provide an effective way to determine linear models for such processes. It has become a standard practice in chemical process control, as can be seen in recent textbooks in process control [3,4]. Wang et al. [5] discuss various aspects of the relay feedback.

The distinct advantages of the relay feedback are:

1. It identifies process information around the important frequency, the ultimate frequency (the frequency where the phase angle is $-\tau$).

2. It is a closed-loop test; therefore, the process will not drift away from the nominal operating point.

3. For processes with a long time constant, it is a more time-efficient method than conventional step or pulse testing. The experimental time is roughly equal to two to four times the ultimate period.

3.1 Experimental Design

Consider a relay feedback system where $G(s)$ is the process transfer function, y is the controlled output, y^{set} is the SP, e is the error and u is the manipulated input (Figure 3.1A).

An on–off (ideal) relay is placed in the feedback loop. The Åström–Hägglund relay feedback system is based on the observation: when the output lags behind the input by $-\tau$ radians, the closed-loop system may oscillate with a period P_u. Figure 3.1(B) illustrates how the relay feedback system works. A relay of magnitude h is inserted in the feedback loop. Initially, the input u is increased by h. As the output y starts to increase (after a dead time D), the relay switches to the opposite position, $u = -h$. Since the phase lag is $-\tau$, a limit cycle with a period P_u results (Figure 3.1). The period of the limit cycle is the ultimate period. Therefore, the ultimate frequency from this relay feedback experiment is

$$\omega_u = \frac{2\tau}{P_u} \qquad (3.1)$$

From the Fourier series expansion, the amplitude a can be considered to be the result of the primary harmonic of the relay output. Therefore, the ultimate gain can be approximated as [1,6]

$$K_u = \frac{4h}{\tau a} \qquad (3.2)$$

where h is the height of the relay and a is the amplitude of oscillation. These two values can be used directly to find controller settings. Notice that Equations 3.1

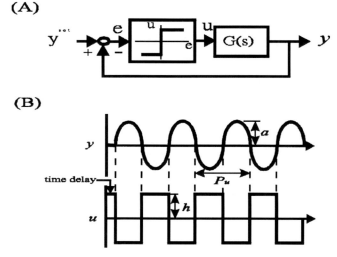

Figure 3.1. (A) Block diagram for a relay feedback system and (B) relay feedback test for a system with positive steady state gain

and 3.2 give approximate values of ω_u and K_u. A more accurate expression will be derived shortly.

The relay feedback test can be carried out manually (without any autotuner). The procedure requires the following steps.

1. Bring the system to steady state.

2. Make a small (*e.g.* 5%) increase in the manipulated input. The magnitude of change depends on the process sensitivities and allowable deviations in the controlled output. Typical values are between 3 and 10%.

3. As soon as the output crosses the SP, the manipulated input is switched to the opposite position (*e.g.* –5% change from the original value).

4. Repeat step 2 until sustained oscillation is observed (Figure 3.1).

5. Read off ultimate period P_u from the cycling and compute K_u from Equation 3.2.

This procedure is relatively simple and efficient. Physically, it implies moving the manipulated input *against* the process. Consider a system with a positive steady state gain (Figure 3.1). When you increase the input (as in step 1), the output y tends to increase also. As a change in the output is observed, you switch the input to the opposite direction. This is meant to bring the output back down to the SP. However, as soon as the output comes down to the SP, you switch the input to the upper position. Consequently, a continuous cycling results, but the amplitude of oscillation is under your control (by adjusting h). More importantly, in most cases, you obtain the information you need for tuning of the controller.

Several characteristics can be seen from the relay feedback test. Consider the most common FOPDT systems.

$$G(s) = \frac{K_p e^{-Ds}}{\tau s + 1} \tag{3.3}$$

where K_p is the steady state gain, D is the dead time and τ is the time constant. Figure 3.2 indicates that, if the normalized dead time D/τ is less than 0.28, the ultimate period is smaller than the process time constant. In terms of plant test, that implies the relay feedback test is more time efficient than the step test. The reason is that it takes almost 3τ to reach 95% of the steady state value in a step test and the time required for the relay feedback is also roughly equal to $3P_u$ (to establish a stable oscillation). Therefore, the relay feedback system is more time efficient than the step test for systems with

$$D/\tau < 0.28 \tag{3.4}$$

Since the dead time cannot be too large (it often comes from the measurement delay), the temperature and composition loops in process industries seem to fall into this category. In other words, Equation 3.4 is fairly typical for many slow chemical processes, especially for units involved with composition changes.

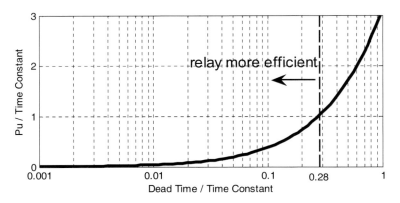

Figure 3.2. P_u/τ as function of the normalized dead time D/τ

3.2 Approximate Transfer Functions: Frequency-domain Modeling

After the relay feedback experiment, the estimated ultimate gain $\hat{K}_u$ and ultimate frequency $\hat{\omega}_u$ can be used directly to calculate controller parameters. Alternatively, it is possible to back-calculate the approximated process transfer functions. The other data useful in finding the transfer function are the dead time D and/or the steady state gain K_p.

In theory, the steady state gain can be obtained from plant data. One simple way to find K_p is to compare the input and output values at two different steady states. That is:

$$K_p = \Delta y / \Delta u \qquad (3.5)$$

where Δy denotes the change in the controlled variable and Δu stands for the deviation in the manipulated input. However, precautions must be taken to make sure that the sizes of the changes in u are made small enough such that the gain in Equation 3.5 truly represents the linearized gain. For highly nonlinear processes, these changes are typically as small as 10^{-3} to 10^{-6} % of the full range [2]. Such small changes would only be feasible using a mathematical model. Trying to obtain reliable steady state gains from plant data is usually impractical.

The dead time D in the transfer function can be easily read off from the initial part of the relay feedback test. It is simply the time it takes for y to start responding to the change in u (Figure 3.1). For the FOPDT system, it is simply the time to reach the peak amplitude in a half period, as will be shown in Chapter 4. Therefore, it is more likely that we will have information on the dead time rather than the steady state gain.

Now we are ready to find an approximate model. Typical transfer functions in process control are assumed and parameters can be calculated. The transfer functions have the following forms:

Model I (integrator plus dead time)

$$G(s) = \frac{K_p e^{-Ds}}{s} \tag{3.6}$$

Model P (pure dead time)

$$G(s) = K_p e^{-Ds} \tag{3.7}$$

Model 1 (FOPDT)

$$G(s) = \frac{K_p e^{-Ds}}{\tau s + 1} \tag{3.8}$$

Model 2a (second-order plus dead time)

$$G(s) = \frac{K_p e^{-Ds}}{(\tau s + 1)^2} \tag{3.9}$$

Model 2b (second-order plus dead time with two unequal lags)

$$G(s) = \frac{K_p e^{-Ds}}{(\tau_1 s + 1)(\tau_2 s + 1)} \tag{3.10}$$

In these five models, model I and model P have two unknown parameters, models 1 and 2a have three unknown parameters and model 2b has four unknown parameters. Therefore, additional information, such as D or K_p, is needed if the last three models are employed. As pointed out by Tyreus and Luyben [7], the simplest integrator-plus-time-delay model (model I) provides good approximation for slow chemical processes, *e.g.* systems showing a small D/τ value. It is the model we recommend for slow processes.

The relay feedback experiment has the following steps:

1. If necessary, the dead time D can be read off from the initial response, or the time to the peak amplitude, and the steady state gain can be obtained from steady state simulation.
2. The ultimate gain $\hat{K}_u$ and ultimate frequency $\hat{\omega}_u$ are computed (Equations 3.1 and 3.2) after the relay feedback experiment.
3. Different model structures (Equations 3.6–3.10) are fitted to the data.

3.2.1 Simple Approach

Once the model is selected, we can back-calculate the model parameters from two equations describing the ultimate gain and the ultimate frequency.

Model I (Friman and Waller [8])

$$K_p = \frac{\omega_u}{K_u} = \frac{2\tau}{K_u P_u} \qquad (3.11)$$

$$D = \frac{\tau}{2\omega_u} = \frac{P_u}{4} \qquad (3.12)$$

Notice that no *a priori* process knowledge is needed for this model. Moreover, computation of K_p and D is quite straightforward.

Model P

$$K_P = \frac{1}{K_u} \qquad (3.13)$$

$$D = \frac{P_u}{2} \qquad (3.14)$$

Similar to model I, no *a priori* process knowledge is necessary.

Model 1

$$\tau = \frac{\tan(\tau - D\omega_u)}{\omega_u} \qquad (3.15)$$

$$\tau = \frac{\sqrt{(K_p K_u)^2 - 1}}{\omega_u} \qquad (3.16)$$

For model 1, either D or K_p is needed to solve for the time constant. For example, if the dead time is read off from the relay test, then we can compute τ from Equation 3.15. Then, K_p can be found by solving Equation 3.16.

Model 2a

$$\tau = \frac{\tan(\tau - D\omega_u)/2}{\omega_u} \qquad (3.17)$$

$$\tau = \frac{\sqrt{(K_p K_u) - 1}}{\omega_u} \qquad (3.18)$$

The equations describing model 2a are quite similar to those for model 1. Again, we need to know D or K_p before finding model parameters.

Model 2b

$$-\tau = -\omega_u D - \tan^{-1}(\omega_u \tau_1) - \tan^{-1}(\omega_u \tau_2) \qquad (3.19)$$

$$\frac{1}{\hat{K}_u} = \frac{K_p}{\sqrt{\left[1+(\omega_u \tau_1)^2\right]\left[1+(\omega_u \tau_2)^2\right]}} \qquad (3.20)$$

Since we have four parameters in model 2b, both K_p and D have to be known in order to solve for the two time constants τ_1 and τ_2. This is the most complex model structure in our models, and it is often sufficient for process control applications.

Let us use an FOPDT system to illustrate the parameter estimation procedure.

Example 3.1 WB column [9]

$$G(s) = \frac{12.8 e^{-s}}{16.8s + 1}$$

This is the transfer function between the top composition x_D and the reflux flow R. From a relay feedback test, we obtain the following ultimate gain and ultimate frequency: $\hat{K}_u = 1.71$ and $\hat{\omega}_u = 1.615$. Note that these two values are only an approximation to the true values: $K_u = 2.1$ and $\omega_u = 1.608$.

Parameters can be calculated for different model structures:

Model I (no prior knowledge on K_p and D)

$$G(s) = \frac{0.94 \, e^{-0.97s}}{s}$$

Model P (no prior knowledge on K_p and D)

$$G(s) = 0.58 e^{-1.94s}$$

Model 1 (assume D is known, i.e. $D = 1$)

$$G(s) = \frac{13.2 \, e^{-s}}{(14.0s + 1)}$$

Model 2a (assume D is known)

$$G(s) = \frac{1.12 \, e^{-s}}{(0.59s + 1)^2}$$

Model 2b (assume K_p and D are known)

$$G(s) = \frac{12.8 e^{-s}}{(13.5s + 1)(0.0009s + 1)}$$

Despite varying in model parameters, all these four models have the *same* ultimate gain and ultimate frequency. That is, the models are correct around the ultimate frequency, which is important for the controller design. However, if we extrapolate the model to different frequencies, e.g. $\omega = 0$, then the results can be completely misleading. For example, the steady state gain of model 2a is only 1.12, which is less than 10% of the true value. We have to be very cautious when using these models. ∎

3.2.2 Improved Algorithm

In theory, if the model structure is correct and the ultimate gain and ultimate frequency are correctly identified, then we could have a very good approximation of the transfer function. For example, if the K_u and ω_u in the previous example are close to the true values, then we will not have errors in the steady state gains and time constant for model 1. Unfortunately, since Equations 3.1 and 3.2 only give approximations to the ultimate gain and ultimate frequency, the parameters derived from Equations 3.15 and 3.16 can deviate significantly from the true system parameters. This implies the observed ultimate period $\hat{P}_u$ and the computed ultimate gain are not the true values.

In order to have a better approximation of the transfer function, fundamental analysis of the relay feedback system is necessary. First, one would like to know what the period of oscillation from the relay feedback experiment really represents. In other words, given a transfer function with known parameters, what is the expression for the period of oscillation observed from the relay feedback experiment, $\hat{P}_u$? The following theorem [1] provides the answer.

Theorem 3.1 Consider the relay feedback system with a transfer function $G(s)$ and an ideal relay (Figure 3.1). Let $HG(T_s, z)$ be the pulse transfer function of $G(s)$ with a sampling time of T_s. If there is a periodic oscillation, then the period of oscillation $\hat{P}_u$ is given by

$$HG(\hat{P}_u/2, -1) = 0$$

Åström and Hägglund [1] prove the theorem starting form the discrete-time state-space equations. The result, $HG(\hat{P}_u/2, -1) = 0$, is obtained by finding the z-domain equivalent. The continuous-time response of an ideal relay (Figure 3.1) can be discretized at the point when the relay switches. The z-transforms of the input and output are $h/(z+1)$ and 0 respectively. Since this is a self-oscillation system, the propagation of the input is described by the gain $HG(\hat{P}_u/2, -1) = 0$. This equation can be used to find the period of oscillation for a known system. In identification, $\hat{P}_u$ is observed from the response and one is able to use this to back-calculate system parameters. Unlike the continuous-time analysis based on the primary harmonic, the discrete-time expression gives a sound basis for finding the system parameters, since no assumption is made in the derivation.

Based on the theorem, a better relationship between $\hat{\omega}_u$ (or $\hat{P}_u$) and the system parameters can be derived. For the transfer functions of interest (models 1, 2a and 2b), the following results can be derived from the modified z-transform [10]:

Model 1

$$\tau = \frac{\pi}{\hat{\omega}_u \ln(2\exp(D/\tau) - 1)} \quad (3.21)$$

Model 2a

$$\tau = \frac{2\pi\left[m + (m-1)\exp\left(-\frac{\pi}{\tau\hat{\omega}_u}\right)\right]}{\hat{\omega}_u\left[1 + \exp\left(-\frac{\pi}{\tau\hat{\omega}_u}\right)\right]\left[\exp\left(\frac{m\pi}{\tau\hat{\omega}_u}\right)\left(1 + \exp\left(-\frac{\pi}{\tau\hat{\omega}_u}\right)\right) - 2\right]} \quad (3.22)$$

where $m = 1 - \dfrac{D\hat{\omega}_u}{\pi}$

Model 2b

$$\tau_1\left[\frac{2\exp\left(-\frac{m\pi}{\tau_1\hat{\omega}_u}\right)}{1 + \exp\left(-\frac{\pi}{\tau_1\hat{\omega}_u}\right)}\right] - \tau_1 = \tau_2\left[\frac{2\exp\left(-\frac{m\pi}{\tau_2\hat{\omega}_u}\right)}{1 + \exp\left(-\frac{\pi}{\tau_2\hat{\omega}_u}\right)}\right] - \tau_2 \quad (3.23)$$

Equations 3.21–3.23 provide alternative expressions between the observed ultimate period, *e.g.* $\hat{\omega}_u$, and system parameters. For example, Equation 3.21 relates $\hat{\omega}_u$ to D and τ in a way that differs substantially from the standard phase angle equation (*i.e.* Equation 3.15).

$$-\pi = -\hat{\omega}_u D - \tan^{-1}(\hat{\omega}_u \tau)$$

Again, we can derive a better expression for the amplitude ratio part at the ultimate frequency, since the expression in Equation 3.2 is based on the first harmonic of the Fourier series expansion. The square-wave response of u (Figure 3.1) consists of many frequency components:

$$u(t) = \frac{4h}{\pi}\sum_{n=0}^{\infty}\frac{\sin((2n+1)\omega t)}{2n+1} \quad (3.24)$$

Therefore, it becomes obvious that the amplitude observed in the relay feedback response is contributed from multiple frequencies, $\omega = \hat{\omega}, 3\hat{\omega}, 5\hat{\omega}$, etc. In theory, one can have a better estimate of the amplitude ratio by employing more terms. An iterative procedure is necessary if more than one term is employed (*e.g.* finding $G(s)$ from the single-term solution and including the higher frequency information, $\omega = 3\hat{\omega}_u$, to find a new $G(s)$ and the procedure is repeated until $G(s)$ converges). However, experimental results show that the estimation of system parameters can be improved substantially by improving the expression for period of oscillation alone, as shown in the next section. Furthermore, for higher order systems, there is little incentive to improve the expression for the amplitude by including more terms, since higher order harmonics (*e.g.* $\omega = 3\hat{\omega}_u$ or $\omega = 5\hat{\omega}_u$) are attenuated by the process. If only one term is employed, then the equations describing the amplitude ratio are exactly the same as Equations 3.16, 3.18 and 3.20.

3.2.3 Parameter Estimation

From the ongoing analysis, the procedure for the evaluation of the transfer function has the following steps:
1. Select model structure.
2. Compute model parameters according to Table 3.1.

Table 3.1 summarizes the information required and the corresponding equations to find the approximate transfer function. Most of these equation sets can be solved sequentially. Notice that if the improved algorithm is used, then better estimates of the ultimate gain and ultimate frequency can be *calculated* from the *model*. For model 2b, if some information is not known, then a different procedure should be employed. For example, if K_p is not available, we can perform a second relay feedback test [11] or use a biased relay (Chapters 7 and 12) to find additional information. Nonetheless, the equations noted in Table 3.1 are generally applicable regardless of the procedure.

3.2.4 Examples

Several examples are used to illustrate the advantages of the improved algorithm. Consider a first-order plus dead time system.

Example 3.2 FOPDT process

$$G(s) = \frac{16.5e^{-10s}}{20s+1}$$

From a relay feedback experiment with $h = 0.04$ we have $\hat{P}_u = 33.26$ and $a = 0.26$. If D and/or K_p are available, we can back-calculate τ. The τ values calculated from Equations 3.15 and 3.16 are $\tau = 16.3$ and 16.09 respectively. The improved algorithm (Equation 3.21) gives a better estimate in τ, $\tau = 19.97$, by improving the expression in the period of oscillation alone. The result from Equation 3.21 is almost exact (the difference may have resulted from reading off a and P_u from the response curve). Figure 3.3 shows the multiplicative modeling errors, $e_m = |(G(i\omega) - \hat{G}(i\omega))/\hat{G}(i\omega)|$, for the transfer function $\hat{G}$ estimated from Equations 3.15, 3.16 and 3.21. The results show that the error e_m is significantly less when τ is calculated from Equation 3.21 alone. ∎

In the following examples, we assume K_p and D are known and the time constant τ for models 1 and 2a is obtained by taking the average of the values calculated from the corresponding equations for the case of the simple algorithm. Next, the effects of dead time on the estimation of the ultimate gain and ultimate frequency are also investigated. In the original ATV method, $\hat{K}_u$ is calculated from

Table 3.1. Equations for different model structures

Model	Simple algorithm	Improved algorithm	Prior information
Model I	Equations 3.11 and 3.12	–	None
Model P	Equations 3.13 and 3.14	–	None
Model 1	Equations 3.15 and 3.16	Equations 3.21 and 3.16	D or K_p
Model 2a	Equations 3.17 and 3.18	Equations 3.22 and 3.18	D or K_p
Model 2b	Equations 3.19 and 3.20	Equations 3.23 and 3.20	D and K_p

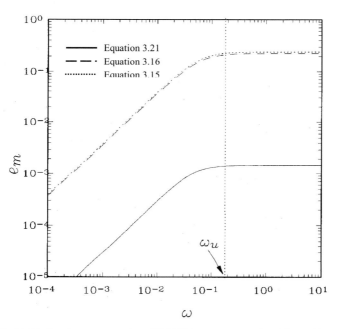

Figure 3.3. Multiplicative errors of an FOPDT system obtained from Equations 3.15, 3.16 and 3.21

Equation 3.2 and $\hat{\omega}_u$ is derived from Equation 3.1. In the proposed method, K_u and ω_u are back-calculated from the estimated transfer function $\hat{G}(s)$. Again, this is shown in the following transfer function:

Example 3.3 Variable dead time

$$G(s) = \frac{16.5e^{-Ds}}{20s+1}$$

The percentage errors in K_u and ω_u are compared for these two methods over a range of dead time (D = 0.1–60). The results (Figure 3.4) show that the errors in K_u for the simple method are quite significant (5–20%). Furthermore, the error in ω_u is almost nil for the improved method. ∎

Similar behavior can also be observed for a second-order lag with time-delay system.

Example 3.4 Second-order system with two unequal lags

$$G(s) = \frac{37.3e^{-Ds}}{(7200s+1)(2s+1)}$$

Figure 3.5 shows that a better estimation of $\hat{G}(s)$ can be achieved over a range of D (D < 60). Again, improvements can be made in finding the correct K_u and ω_u by using a more accurate expression in the period of oscillation. ∎

Since the estimated transfer function is typically employed in the analysis and design of a feedback control system, the *impact* of the modeling errors in closed-loop performance is evaluated. A model-based controller, IMC, is employed to analyze the performance. One of the advantages of the IMC is that we can specify the desired trajectory in the design. Figure 3.6 compares the SP responses of IMC when different models $\hat{G}'$ are employed in the design of the controllers. Consider

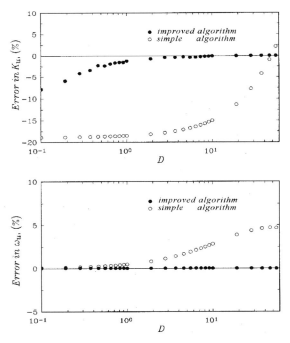

Figure 3.4. Percentage errors in K_u and ω_u for the FOPDT system over a range of dead time D

the FOPDT system

$$G(s) = \frac{16.5e^{-10s}}{20s+1}$$

The SP response of the control system, designed according to $\hat{G}(s)$ from the simple algorithm, tends to be more sluggish than the desired trajectory (Figure 3.6). The proposed method improves this situation, as shown in Figure 3.6. Despite the fact that a tighter response can be achieved by shortening the closed-loop time constant under modeling errors, one has to realize that the value of a model-based controller is that one can foresee the closed-loop response. In other words, a good model always helps.

Generally, the proposed method improves the estimation in $G(s)$ at the nominal condition (with perfect knowledge of K_p and D). The robustness with respect to errors in the dead time is investigated. Since the improved method calculates K_u and ω_u by finding the transfer function $\hat{G}(s)$ first, followed by solving the corresponding equations for them, it is more sensitive to the errors in the dead time than the original method. Let us take another FOPDT system as an example.

Example 3.5 Error in the observed dead time

$$G(s) = \frac{16.5e^{-s}}{20s+1}$$

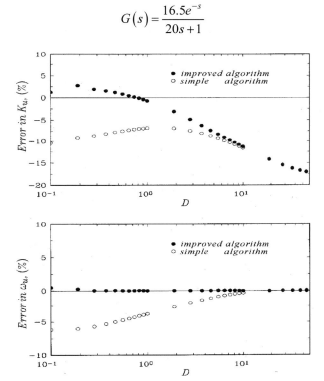

Figure 3.5. Percentage errors in K_u and ω_u for a second-order plus dead time system over a range of delay time D

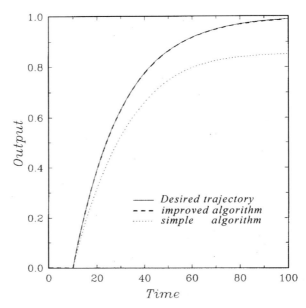

Figure 3.6. SP responses of IMC designed according to the estimated transfer functions $\hat{G}(s)$ (the closed-loop constant is 20 for the desired trajectory)

Figure 3.7 shows the estimate of K_u and ω_u for both methods when the percentage errors in dead time range from –50% to 50%. Despite the fact that the errors in K_u and ω_u are less for the improved method over a reasonable range of errors in dead time, it is more sensitive to the error in D. Therefore, care should be taken in reading off the dead time from the initial responses or the time to the peak amplitude. ■

3.3 Approximate Transfer Functions: Time-domain Modeling

Up to this point, the model identification is based on the frequency domain approach, which is based on the describing functions. A method to derive FOPDT-type systems was proposed by Wang et al. [12] using a single relay test. In a separate attempt, Majhi and Atherton [3] proposed a technique to identify plant parameters, but the method needs a correct initial guess and convergence is not guaranteed. Kaya and Atherton [14] describe another method (A-locus) to identify low-order process parameters from relay autotuning. Panda and Yu [15] develop analytical models to represent relay responses produced by different systems. The relay output consists of a series of step changes in manipulated variables (with opposite sign). Hence, the stabilized output is a sum of infinite terms of step responses due to those step changes. For systems with dead time D, the actual relay output

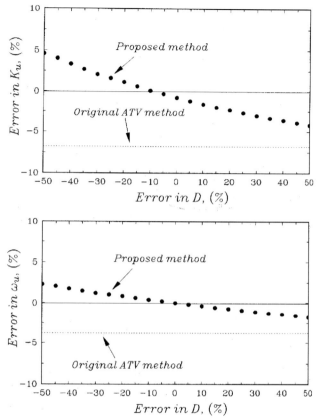

Figure 3.7. Percentage errors in K_u and ω_u for a first-order system over a range of variation in the dead time

lags behind the input by a time unit D. The inputs and outputs can be synchronized by shifting the output forward in time by an amount D, as shown in Figure 3.8B, and, in doing this, the dead time D can be eliminated from the expression for relay responses, as will be shown later. The shifted version of a typical relay feedback response provides the basis for the derivation.

It is assumed that the relay response is formed by n-number of step changes, of opposite directions ($\pm u$), in input. The switching period for each step change is $P_u/2$, except for the initial step change. In Figure 3.9, in the first interval, as time changes from $t = 0$ to $t = D$, the response y_1 is produced due to the first step change u_1. Again, in the second interval, time progressing from D to $D + P_u/2$, response y_2 results due to the combined effects of step changes u_1 and u_2. Similarly, the effect of u_1, u_2 and u_3 produces y_3 during the third time interval ($D + P_u/2$ to $D + P_u$). Two half periods ($P_u/2$) are of special interest in Figure 3.9. The even values of n result in descending half period y_{2n}, and the odd values of n formulate the ascending half periods y_{2n+1}. It is interesting to note that the

38　Autotuning of PID Controllers

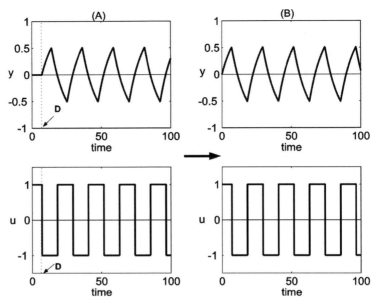

Figure 3.8. Schematic representation of the shifted version of relay feedback response for the development of their analytical expressions: (A) original relay feedback responses and (B) output y shifted by D

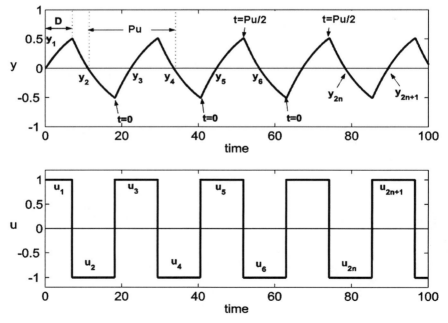

Figure 3.9. Shifted version of relay input u and output y response of a typical SOPDT system

3.3.1 Derivation for a Second-order Overdamped System

The transfer function of an SOPDT system with a damping coefficient greater than one can be expressed as $G(s) = K_p e^{-Ds} / [(\tau_1 s+1)(\tau_2 s+1)]$, where K_p is the steady state gain, τ_1 and τ_2 are process time constants with $\tau_1 > \tau_2$, and D is the dead time. The original step response of an overdamped SOPDT can be given by

$$y = K_p [1 - a_1 e^{-(t-D)/\tau_1} + b_1 e^{-(t-D)/\tau_2}]$$

where a_1 and b_1 are given by

$$a_1 = \frac{\tau_1}{\tau_1 - \tau_2} \quad \text{and} \quad b_1 = \frac{\tau_2}{\tau_1 - \tau_2}$$

Under the shifted version (Figure 3.8B), the first segment of the relay response y_1 is simply the step response without dead time in the time index:

$$y_1 = K_p [1 - a_1 e^{-t/\tau_1} + b_1 e^{-t/\tau_2}] \tag{3.25}$$

At the second instant, the time is reset to zero at the initial point. The step response (relay output) is given by (i.e. introducing a time shift by D amount in Equation 3.22)

$$y_2 = K_p \left[1 - a_1 e^{-\frac{t+D}{\tau_1}} + b_1 e^{-\frac{t+D}{\tau_2}}\right] - 2K_p \left[1 - a_1 e^{-\frac{t}{\tau_1}} + b_1 e^{-\frac{t}{\tau_2}}\right]$$

Here, the first term represents the effect of the first step change (occurred at D time earlier) and the second term shows the effect of the second step input, switching to the opposite direction. The above equation can be simplified to

$$y_2 = K_p \left\{ [1-2] - a_1 e^{-\frac{t}{\tau_1}} \left(e^{-\frac{D}{\tau_1}} - 2\right) + b_1 e^{-\frac{t}{\tau_2}} \left(e^{-\frac{D}{\tau_2}} - 2\right) \right\} \tag{3.26}$$

The relay response at the third interval is the result of three step changes, lags by an amount $D + P_u / 2$ from input. After introducing a time shift of $D + P_u / 2$ in Equation 3.22, the net effect becomes

$$y_3 = K_p \left\{ \left[1 - a_1 e^{-\frac{t+D+P_u/2}{\tau_1}} + b_1 e^{-\frac{t+D+P_u/2}{\tau_2}}\right] \right.$$
$$\left. -2\left[1 - a_1 e^{-\frac{t+P_u/2}{\tau_1}} + b_1 e^{-\frac{t+P_u/2}{\tau_2}}\right] + 2\left[1 - a_1 e^{-\frac{t}{\tau_1}} + b_1 e^{-\frac{t}{\tau_2}}\right] \right\}$$

which can be simplified further as

---

[1] One may skip the derivation in Section 3.3.1 and refer directly to Tables 3.2 and 3.3 for the results.

$$y_3 = K_p \left\{ [1-2+2] - a_1 e^{-\frac{t}{\tau_1}} \times \left[e^{-\frac{D+P_u/2}{\tau_1}} - 2e^{-\frac{P_u}{2\tau_1}} + 2 \right] + b_1 e^{-\frac{t}{\tau_2}} \right.$$
$$\left. \times \left[e^{-\frac{D+P_u/2}{\tau_2}} - 2e^{-\frac{P_u}{2\tau_2}} + 2 \right] \right\} \tag{3.27}$$

It can be seen that the terms in the right-hand side (RHS) of the above equation are slowly forming a series.

With the progress of time, the response becomes stabilized and the general expression for the nth term can be described as

$$y_n = K_p \left\{ [1-2+2-\cdots] - a_1 e^{-\frac{t}{\tau_1}} \left[e^{-\frac{D+(n-2)P_u/2}{\tau_1}} \right. \right.$$
$$\left. -2e^{-\frac{(n-2)P_u}{2\tau_1}} + 2e^{-\frac{(n-1)P_u}{2\tau_1}} - \cdots + 2e^{-\frac{P_u}{2\tau_1}} - 2 \right]$$
$$+ b_1 e^{-\frac{t}{\tau_1}} \left[e^{-\frac{D+(n-2)P_u/2}{\tau_2}} - 2e^{-\frac{(n-2)P_u}{2\tau_2}} + 2e^{-\frac{(n-1)P_u}{2\tau_2}} \right.$$
$$\left.\left. - \cdots + 2e^{-\frac{P_u}{2\tau_2}} - 2 \right] \right\} \tag{3.28}$$

The RHS of Equation 3.28 has three parts, and each part consists of an infinite series, F_1, F_2 and F_3.

$$y_n = K_p \left\{ F_1 - a_1 e^{-\frac{t}{\tau_1}} F_2 + b_1 e^{-\frac{t}{\tau_2}} F_3 \right\}$$

If n is odd, the first series F_1 is simply

$$F_1 = [1 - 2 + 2 - 2 + \cdots] = 1$$

The second series becomes:

$$F_2 = \left[e^{-\frac{D}{\tau_1}} r^{n-2} - 2r^{n-2} + 2r^{n-3} - 2r^{n-4} + \cdots - 2r + 2 \right]$$

where $r = e^{-v_1}$ and $v_1 = P_u/2\tau_1$. This above series is convergent and can be put into the following form (note that terms are rearranged from the back side of the above expression):

$$F_2 = \lim_{n \to \infty} \left(e^{-D/\tau} r^{n-2} \right) + 2\left(1 - r + r^2 - r^3 + \cdots \right)$$
$$= 2\left[1 - r + r^2 - r^3 + \cdots \right] = \frac{2}{1+r} = \frac{2}{1 + e^{-P_u/2\tau_1}}$$

In a similar way, the F_3 of the RHS of Equation 3.28 can also be simplified. Ultimately, the response can be given by

$$y_n = K_p \left\{ 1 - a_1 e^{-\frac{t}{\tau_1}} \left(\frac{2}{1 + e^{-\frac{P_u/2}{\tau_1}}} \right) + b_1 e^{-\frac{t}{\tau_2}} \left(\frac{2}{1 + e^{-\frac{P_u/2}{\tau_2}}} \right) \right\} \quad (3.29)$$

This represents the ascending response (n is odd). Since this response is dissymmetric, the general form can be employed as

$$y_n = K_p \left\{ -1 + a_1 e^{-\frac{t}{\tau_1}} \left(\frac{2}{1 + e^{-\frac{P_u/2}{\tau_1}}} \right) - b_1 e^{-\frac{t}{\tau_2}} \left(\frac{2}{1 + e^{-\frac{P_u/2}{\tau_2}}} \right) \right\} (-1)^n \quad (3.30)$$

One can refer to Panda and Yu [15] for the derivations for critically damped and underdamped SOPDT systems, as well as for high-order systems.

3.3.2 Results

Different types of transfer function are considered, and the analytical expressions for their relay feedback output response are developed following the above procedure. Table 3.2 gives a list of first-, second-, and third-order plus dead time processes and their corresponding mathematical expressions for the stabilized relay feedback output responses. These equations y_n denote the upward or ascending trend (or sometimes, curves in the lower part of midline for higher order systems) of relay feedback output (while time t changes from 0 to $P_u/2$). The downward or descending trend can be obtained by reversing the sign of the output ($-y_n$).

In Table 3.2, the individual expressions, for relay feedback responses of first-, second- and third-order systems contain terms similar to those of the corresponding equations for the step responses, except that they differ only in weighting factor ($2/(1+e^{-P_u/2\tau})$). If we compare the terms of the expressions of the relay feedback response with those of step response of a process, we see that they differ by a weighting factor of $2/(1+e^{-P_u/2\tau})$. For an FOPDT system, the response starts ($t=0$) from the minimal point, at $y=-a$, and ends ($t=P_u/2$) at the maximal point, at $y=a$. Also note that, for an unstable FOPDT system, stable limit cycles can occur only if $D/\tau < \ln(2)$. For the lead/lag second-order system (No. 6 in Table 3.2), the expression is applicable to systems with left-half plane ($\tau_3 > 0$) or right-half plane ($\tau_3 < 0$) zero.

Analytical expressions of relay feedback output responses for higher order systems are presented in Table 3.3. They are of much interest because, when we see, for example, the expression for fifth-order process, the equation contains mainly five terms (except '1') and each of these terms represents corresponding lower order processes. The first term inside the third bracket of the first line/row appears to be for an FOPDT. The second term (having two terms inside the first bracket) is for an SOPDT (critically damped). The third term (having three terms inside the first bracket) is for a third-order process. The terms in the second row/line (having

Table 3.2. Time response y_n of relay feedback for FOPDT, SOPDT and third-order processes (y_n is for ascending part of response; $-y_n$ is for descending part)

No	Process Transfer Functions	Time response (y_n) of Relay Feedback
1	$\dfrac{K_p e^{-Ds}}{\tau s+1}$	$K_p\left(1-e^{-t/\tau}\left[\dfrac{2}{1+e^{-P_u/2\tau}}\right]\right)$
2	$\dfrac{K_p e^{-Ds}}{\tau s-1}$	$K_p\left(1-e^{t/\tau}\left[\dfrac{2}{1+e^{P_u/2\tau}}\right]\right)$. $(D/\tau < \ln(2))$
3	$\dfrac{K_p e^{-Ds}}{(\tau s+1)^2}$	$K_p\left\{1-e^{-\frac{t}{\tau}}\left[\dfrac{2}{1+e^{-P_u/2\tau}}\right] - 2e^{-\frac{t}{\tau}}\left[\dfrac{t/\tau}{1+e^{-P_u/2\tau}} + \dfrac{-(P_u/2\tau)e^{-P_u/2\tau}}{(1+e^{-P_u/2\tau})^2}\right]\right\}$
4	$\dfrac{K_p e^{-Ds}}{(\tau_1 s+1)(\tau_2 s+1)}$	$K_p\left\{1-a_1 e^{-\frac{t}{\tau_1}}\left(\dfrac{2}{1+e^{-P_u/2\tau_1}}\right)+b_1 e^{-\frac{t}{\tau_2}}\left(\dfrac{2}{1+e^{-P_u/2\tau_1}}\right)\right\}$. where $a_1=\dfrac{\tau_1}{\tau_1-\tau_2}$ and $b_1=\dfrac{\tau_2}{\tau_1-\tau_2}$ $(\tau_1>\tau_2)$
5	$\dfrac{K_p e^{-Ds}}{\tau^2 s^2+2\xi\tau s+1}$	$K_p\left\{1-2\dfrac{e^{-\frac{\xi t}{\tau}}}{\beta}\sin\left(\dfrac{\beta t}{\tau}+\alpha\right)\right\}$. where $\alpha=\tan^{-1}\left(\dfrac{\beta+\beta r\cos(\theta)-\xi r\sin(\theta)}{\xi+\xi r\cos(\theta)+\beta r\sin(\theta)}\right)$, $\beta=\sqrt{1-\xi^2}$, $r=e^{-\frac{P_u \xi \tau}{2}}$ and $\theta=P_u\cdot\beta\cdot\tau/2$
6	$\dfrac{K_p(\tau_3 s+1)e^{-Ds}}{(\tau_1 s+1)(\tau_2 s+1)}$	$K_p\left\{1-a_1 e^{-\frac{t}{\tau_1}}\left(\dfrac{2}{1+e^{-P_u/2\tau_1}}\right)+b_1 e^{-\frac{t}{\tau_2}}\left(\dfrac{2}{1+e^{-P_u/2\tau_1}}\right)\right\}$. where $a_1=\dfrac{\tau_1-\tau_3}{\tau_1-\tau_2}$ and $b_1=\dfrac{\tau_2-\tau_3}{\tau_1-\tau_2}$ $(\tau_1>\tau_2$; $\tau_3>$ or $<0)$
7	$\dfrac{K_p}{(\tau s+1)^3}$	$K_p\left\{1-2e^{-t/\tau}\left[\left(\dfrac{1}{1-r}\right)+\left(\dfrac{q}{1-r}+\dfrac{rv}{(1-r)^2}\right)+\dfrac{1}{2!}\left(\dfrac{q^2}{1-r}+\dfrac{2qvr}{(1-r)^2}+\dfrac{v^2 r(1+r)}{(1-r)^3}\right)\right]\right\}$ where $q=t/\tau$, $v=P_u/2\tau$, and $r=-e^{-P_u/2\tau}$
8	$\dfrac{K_p}{(\tau_1 s+1)(\tau_2 s+1)(\tau_3 s+1)}$	$K_p\left\{1+a_1 e^{-\frac{t}{\tau_1}}\left(\dfrac{2}{1+e^{-P_u/2\tau_1}}\right)+b_1 e^{-\frac{t}{\tau_2}}\left(\dfrac{2}{1+e^{-P_u/2\tau_1}}\right)+c_1 e^{-\frac{t}{\tau_3}}\left(\dfrac{2}{1+e^{-P_u/2\tau_3}}\right)\right\}$. where $a_1=\dfrac{\tau_1^2(\tau_1-\tau_3)}{(\tau_1-\tau_3)}$, $b_1=\dfrac{\tau_2^2(\tau_3-\tau_2)}{(\tau_1-\tau_2)}$ and $c_1=\dfrac{\tau_3^2}{(\tau_1\tau_2-\tau_2\tau_3-\tau_3\tau_1+\tau_3^2)}$ $(\tau_1>\tau_2>\tau_3)$

Table 3.3. Time response y_n of relay feedback for fourth and high-order processes

No	Process Transfer Functions	Time response (y_n) of Relay Feedback (y_n is for ascending part of response; $-y_n$ is for descending part)
9	$\dfrac{K_p}{(\tau s+1)^4}$	$K_p\left\{1-2e^{-t/\tau}\left[\dfrac{1}{1-r}+\dfrac{q}{1-r}+\dfrac{rv}{(1-r)^2}+\dfrac{1}{2!}\left(\dfrac{q^2}{1-r}+\dfrac{2qvr}{(1-r)^2}+\dfrac{v^2r(1+r)}{(1-r)^3}\right)\right]-\dfrac{2}{3!}e^{-t/\tau}\left[\dfrac{q^3}{1-r}+\dfrac{3q^2vr}{(1-r)^2}+\dfrac{3qv^2r(1+r)}{(1-r)^3}+\dfrac{v^3r(r^2+4r+1)}{(1-r)^4}\right]\right\}$ where $q=t/\tau$, $v=P_u/2\tau$, and $r=-e^{-P_u/2\tau}$
10	$\dfrac{K_p}{(\tau s+1)^5}$	$K_p\left\{1-2e^{-t/\tau}\left[\dfrac{1}{1-r}+\left(\dfrac{q}{1-r}+\dfrac{rv}{(1-r)^2}\right)+\dfrac{1}{2!}\left(\dfrac{q^2}{1-r}+\dfrac{2qvr}{(1-r)^2}+\dfrac{v^2r(1+r)}{(1-r)^3}\right)\right]\right.$ $-\dfrac{2}{3!}e^{-t/\tau}\left[\dfrac{q^3}{1-r}+\dfrac{3q^2vr}{(1-r)^2}+\dfrac{3qv^2r(1+r)}{(1-r)^3}+\dfrac{v^3r(r^2+4r+1)}{(1-r)^4}\right]$ $\left.-\dfrac{2}{4!}e^{-t/\tau}\left[\dfrac{q^4}{1-r}+\dfrac{4q^3vr}{(1-r)^2}+\dfrac{6q^2v^2r(1+r)}{(1-r)^3}+\dfrac{4qv^3r(r^2+4r+1)}{(1-r)^4}+\dfrac{v^4r(r^3+11r^2+11r+1)}{(1-r)^5}\right]\right\}$ where $q=t/\tau$, $v=P_u/2\tau$ and $r=-e^{-P_u/2\tau}$
11	$\dfrac{K_p}{(\tau s+1)^n}$	$K_p\left\{1-2e^{-q}\left[\dfrac{1}{0!}\left(\dfrac{1}{1-r}\right)+\dfrac{1}{1!}\left(\dfrac{q}{1-r}+\dfrac{rv}{(1-r)^2}\right)+\dfrac{1}{2!}\left(\dfrac{q^2}{1-r}+\dfrac{2qvr}{(1-r)^2}+\dfrac{v^2r(1+r)}{(1-r)^3}\right)\right]\right.$ $-\dfrac{2}{3!}e^{-q}\left[\dfrac{q^3}{1-r}+\dfrac{3q^2vr}{(1-r)^2}+\dfrac{3qv^2r(1+r)}{(1-r)^3}+\dfrac{v^3r(r^2+4r+1)}{(1-r)^4}\right]+\ldots$ $\left.-\dfrac{2}{(n-1)!}e^{-q}\left[\dfrac{^{n-1}C_0q^{n-1}}{1-r}+\dfrac{^{n-1}C_1q^{n-2}vr}{(1-r)^2}+\dfrac{^{n-1}C_2q^{n-3}v^2r(1+r)}{(1-r)^3}+\dfrac{^{n-1}C_3qv^3r(r^2+4r+1)}{(1-r)^4}+\ldots+{}^{n-1}C_{n-1}v^{n-1}\sum_{d=0}^{\infty}d^{n-1}r^d\right]\right\}$ where $q=t/\tau$, $v=P_u/2\tau$, and $r=-e^{-P_u/2\tau}$

four terms inside) are for a fourth-order process. In the third or last row/line there are five terms for a fifth-order process. Hence, the number of terms (size of the series) for a particular order of process is rhythmic. These tables are similar to the tables of inverse Laplace transform and will help in finding an equation for relay feedback responses.

3.3.3 Validation

Two kinds of response can be observed in the analytical expressions in Tables 3.2 and 3.3. These responses are tabulated in Figure 3.10. Systems with serial numbers 1 and 2 in Table 3.2 always produce a monotonic response, where, at $t = 0$, the response from the model starts at the lowermost (or uppermost) point (A or B) and, at $t = P_u/2$, it ends at the other extreme point (B or C). Processes with serial numbers 3, 4, 5 and 6 in Table 3.2 may give a non-monotonic response, as shown in Figure 3.10. The third type is higher order systems without dead time (*i.e.* $n \geq 3$). For this type of system, this value occurs at the mid-point of the half period, as also shown in Figure 3.10.

Figure 3.10 shows the correctness of the derived mathematical models. If the relay height is other than unity, then the model for the relay output response will be just multiplied by actual value of relay height h.

3.4 Conclusion

In this chapter the relay feedback test is introduced and the steps required to perform the experiment are also given. It can be carried out with or without a commercial autotuner. Once you have obtained the information on the ultimate frequency, the controller settings can be decided using the original or modified Ziegler–Nichols methods. You can also go a step further to find an appropriate transfer function for the process. This can be useful for implementing MPC or dead time compensator (Smith predictor). Better approximation can be achieved using the improved algorithm. Finding transfer functions using the biased relay plus hysteresis was discussed by Wang *et al.* [12]. Finally, analytical expressions for relay feedback responses are tabulated for different types of process. This can be useful if the model structure is known.

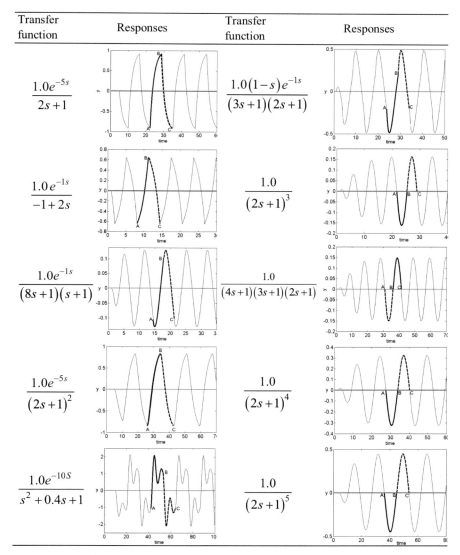

Figure 3.10. Validation of analytical expressions for relay output of different systems: solid line is relay output and dashed line is model output. (A denotes starting of one cycle that ends at B. Again from B next cycle starts and ends at C).

3.5 References

1. Åström KJ, Hägglund T. Automatic tuning of simple regulators with specifications on phase and amplitude margins. Automatica 1984;20:645.
2. Luyben WL. Derivation of transfer functions for highly nonlinear distillation columns. Ind. Eng. Chem. Res. 1987;26:2490.
3. Seborg DE, Edgar TF, Mellichamp DA. Process dynamics and control. 2nd ed. New York: Wiley; 2004.
4. Luyben WL, Luyben ML. Essentials of process control. New York: McGraw-Hill; 1997.
5. Wang QG, Lee TH, Lin C. Relay feedback. London: Springer-Verlag; 2003.
6. Ogata K. Modern control engineering. Prentice-Hall: Englewood Cliffs; 1970.
7. Tyreus BD, Luyben WL. Tuning PI controllers for integrator/dead time processes. Ind. Eng. Chem. Res. 1992;31:2625.
8. Friman M, Waller KV. Autotuning of multiloop control systems. Ind. Eng. Chem. Res. 1994;33:1708.
9. Wood RK, Berry MW. Terminal composition control of a binary distillation column. Chem. Eng. Sci. 1973;28:1707.
10. Chang RC, Shen SH, Yu CC. Derivation of transfer function from relay feedback systems. Ind. Eng. Chem. Res. 1992;31:855.
11. Li W, Eskinat E, Luyben WL. An improved autotune identification method. Ind. Eng. Chem. Res. 1991;30:1530.
12. Wang QG, Hang CC, Zou B. Low-order modeling from relay feedback. Ind. Eng. Chem. Res. 1997;36:375.
13. Majhi S, Atherton DP. Auto-tuning and controller design for processes with small time delays. IEE Proc. Control Theory Appl. 1999;146(3):415.
14. Kaya I, Atherton DP. Parameter estimation from relay auto-tuning with asymmetric limit cycle data. Process Control 2001;11:429.
15. Panda RC, Yu CC. Analytical expressions for relay feedback responses. J. Process Control 2003;13:48.

4
Shape of Relay

Luyben [1] pointed out that the shapes of the response curves of a relay feedback test contain useful information. A simple characterization factor was proposed to quantify the curve shape and later used to determine the three parameters for FOPDT processes. This concept offers an attractive alternative to improve the relay feedback autotuning, because qualitative information of model structure is available.

Here, we intend to utilize the shape information from the relay feedback test to identify the correct model structure of the process and to find appropriate PID controller settings. The additional shape information is also useful to devise dead time compensation and high-order compensation, when necessary. Hieroglyphic writing can often be seen in ancient cultures. Figure 4.1 shows that much of Chinese is written in pictorial characters. The "shapes" of the characters tell us something about their meaning. We intend to extract some useful information from the "shape" of the relay response.

4.1 Shapes of Relay Response

The Åström and Hägglund [2] relay feedback test is a useful tool in identification because it identifies two important parameters, ultimate gain and ultimate frequency, for controller tuning. Typically, the Ziegler–Nichols type of tuning rule is applied because K_u and P_u are the information required to set PID controller parameters. Unfortunately, satisfactory performance is not always guaranteed because no single tuning rule works well for the entire dead time D to time constant τ ratio D/τ even for an FOPDT process. Luyben demonstrates that, for FOPDT processes, a different D/τ ratio gives different shapes in relay feedback tests (Figure 4.2) and this shape factor can be utilized to find the D/τ value and different tuning rules can be applied accordingly. This presents a significant progress in relay feedback identification, and much reliable autotuning has resulted, as shown by Luyben. Figure 4.2 shows the transition from a triangle to an almost rectangular curve as D/τ changes from 0.1 to 10. Similar figures were also given by

48 Autotuning of PID Controllers

Figure 4.1. Hieroglyphic writing of Chinese characters

Friman and Waller [3]. In Luyben's work, time to the mid-point of the amplitude a is used to characterize D/τ.

4.1.1 Shapes

To characterize model structure and parameter value (*e.g.* D/τ), processes with different order (first, second, third, eighth, fifteenth and twentieth order) and dead time to time constant ratio (*i.e.* $D/\tau = 0.01$, 1 and 10) are studied. In this work, only overdamped processes are studied (underdamped processes and systems with inverse response are not included). Figure 4.3 shows the relay feedback responses for those higher order processes. Note that all process gains are assumed to be one and a relay height $h = 1$ is used to generate sustained oscillations.

From the curve shapes, Figures 4.2 and 4.3, several observations can be made immediately.

1. *FOPDT process.*
   If the response curves show a sharp edge (discontinuity) at the peak amplitudes (*i.e.* $y = \pm a$), then the process can be considered as an FOPDT system, as shown in Figure 4.2.

2. *Effect of D/τ for FOPDT process.*
   If the relay feedback gives a triangular wave, then the process can be treated as a time-constant-dominant process (*i.e.* small D/τ for FOPDT). Specifically, the time to reach the peak amplitude is equal to the dead time, as will be shown later. If the dead time to time constant ratio becomes larger, then curvature begins to appear (*e.g.* Figure 4.2), and this implies a gradually developing step response. As D/τ approaches infinity, the response resembles a symmetrical rectangular wave. Actually, FOPDT processes represent a very unique class in terms of relay feedback responses.

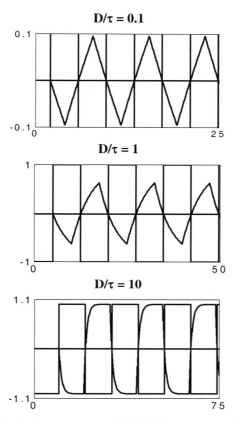

Figure 4.2. Relay feedback responses of FOPDT processes with different D/τ values (controlled variable, solid lines; manipulated variable, dashed lines)

3. *Effect of order.*
   If the order of the process increases to two and beyond (*e.g.* n = 2, 3, 8, 15 and 20 in Figure 4.3), the sharp edge disappears and the responses resemble a sinusoidal oscillation. Generally, a sustained oscillation is developed in the cycles except for the second-order process with small D/τ value (Figure 4.3). Again, when the dead time to time constant ratios become large, the responses approach rectangular waves.

4. *Exponentially developed cycling.*
   If the response is of sinusoidal oscillations with exponentially increasing magnitude and reaching a steady state after many cycles, the process can be considered as a second-order process with small D/τ value. This again represents a special class in relay feedback responses.

50   Autotuning of PID Controllers

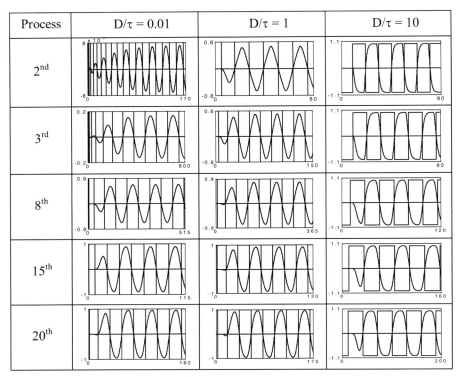

Figure 4.3. Relay feedback response for processes with different orders and various D/τ ratios (controlled variable, solid lines; manipulated variable, dashed lines)

4.1.2 Model Structures

The observations presented above are useful in identifying different model structures. The basic principle in classification is to use the least classes while capturing all possible curve shapes in relay feedback responses. Based on the responses in Figures 4.2 and 4.3, three distinct classes are identified.

4.1.2.1 First-order Plus Dead Time

As pointed out in the previous section, two distinct features constitute FOPDT systems: (1) a response curve showing sharp edges and (2) a response reaching a stationary oscillation in the first cycle. Therefore, category 1 is represented by the FOPDT process

$$G(s) = \frac{K_p e^{-Ds}}{\tau s + 1} \qquad (4.1)$$

where K_p is the steady state gain, D is the dead time and τ denotes the time constant. Figure 4.4 shows that, if the process is truly FOPDT, it certainly falls into category 1. But, a high-order process with a large D/τ value can also be classified into this category, at least by inspecting the relay feedback response, as shown in the top two rows of Figure 4.3 with $D/\tau = 10$. Quantitative comparison of multiplicative error also reveals that high-order systems with $D/\tau > 10$ are better represented by an FOPDT system (Figure 4.4).

4.1.2.2 Second-order Plus Small Dead Time

Observation from Figure 4.3 indicates that we need a model structure to describe exponentially developed cycling (top row with $D/\tau = 0.01$ in Figure 4.3).

An ideal candidate is a second-order plus small dead time (SOPSDT) process:

$$G(s) = \frac{K_p e^{-Ds}}{(\tau s + 1)^2} \tag{4.2}$$

Typically, if the ratio $\varepsilon = D/\tau$ is less than 0.01, then the oscillation develops slowly. In this work, the ratio ε is set to 0.001. Thus, the transfer function can be expressed as

$$G(s) = \frac{K_p e^{-\varepsilon \tau s}}{(\tau s + 1)^2} \tag{4.3}$$

Again, quantitative assessment (Figure 4.4) also confirms such a category, SOPSDT. This is denoted as category 2.

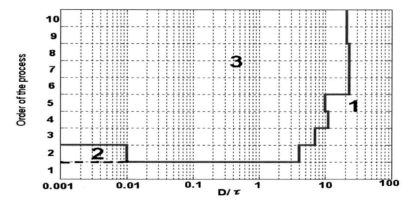

Figure 4.4. Quantitative classification of different model structures (categories 1 to 3) based on integrated absolute error from frequency response

4.1.2.3 High Order

In addition to the two above-mentioned categories, Figure 4.3 indicates that rest of the responses showing a sinusoidal oscillation and stationary cycling is reached in one or two cycles. This behavior can be described by a high-order (HO) process without dead time, which is also called category 3. A typical transfer function is

$$G(s) = \frac{K_p}{(\tau s + 1)^n} \tag{4.4}$$

As long as $n \geq 3$ we can see similar relay feedback responses. In this work, a default value of $n = 5$ is used. This leads to

$$G(s) = \frac{K_p}{(\tau s + 1)^5} \tag{4.5}$$

Figure 4.4 shows that this category, category 3, covers the largest parameter space of the systems studied.

The model structures presented above (Equations 4.2, 4.3 and 4.5) show that there are only two unknown parameters for categories 2 and 3, and for the FOPDT model we have three unknown parameters (*i.e.* K_p, D, and τ). The parameters chosen for categories 2 and 3 (*i.e.* ε and n in Equations 4.3 and 4.4) may affect the distribution of model structure in the parameter space of Figure 4.4. Nonetheless, most of the curve shapes are well represented using these three classes.

4.2 Identification

The detailed procedures for the system identification of various processes under different categories are presented below.

4.2.1 Identification of Category 1: First-order Plus Dead Time

This category includes two types of system. One is the true FOPDT process, as shown in the last row of Figure 4.4, and this is denoted as category 1a. The second one is high order systems with a large D/τ value, as shown in the RHS of Figure 4.4 and it is called category 1b.

4.2.1.1 Category 1a: True First-order Plus Dead Time

This category has two important characteristics: (1) showing a sharp edge at the peak amplitude, and (2) developing a stationery oscillation in the first cycle. The relay feedback response of an FOPDT system actually can be described analytically. Figure 4.5A shows the original response curve where the output starts to increase after the dead time D. If we align the output y with the input u by shifting

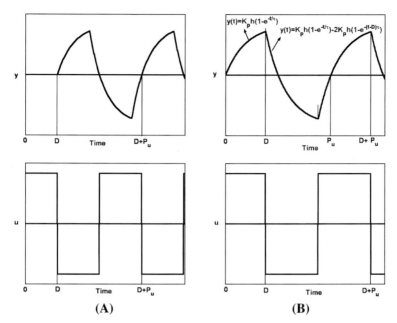

Figure 4.5. Analytical expressions of relay feedback response for FOPDT systems: (A) original response and (B) shifted version

the output to the left (Figure 4.5B), it becomes clear that the increasing part of the output response is the result of a step increase in "u" by a magnitude "h". After a delay time D, the relay switches to $-h$ and the second step change becomes effective immediately. This results in the decreasing portion of in the half cycle shown in Figure 4.5B. Therefore, the analytical expression for the first half cycle becomes

$$y(t) = K_p h(1 - e^{-t/\tau}) \quad \text{for} \quad 0 < t < D \quad (4.6)$$

$$y(t) = K_p h(1 - e^{-t/\tau}) - 2K_p h(1 - e^{-(t-D)/\tau}) \quad \text{for} \quad D < t < P_u/2 \quad (4.7)$$

The continuous step change repeats itself and a sustained oscillation results. Note that similar derivations were proposed by Wang *et al.* [4] for stable FOPDT systems and by Tan *et al.* [5] and Huang and Chen [6] for first-order open-loop unstable systems. Equations 4.6 and 4.7 clearly indicate that the time to reach the peak amplitude is exactly the dead time D for the FOPDT process, and this value can be validated repeatedly at each half cycle ($P_u/2$). Provided with the two boundary conditions $y(D) = a$ and $y(P_u/2) = 0$, we are able to solve for the other two model parameters, K_p and τ:

$$\tau = \frac{P_u/2}{\ln(2e^{D/\tau} - 1)} \quad (4.8)$$

$$K_p = \frac{a}{h(1-e^{-D/\tau})} \tag{4.9}$$

Thus, all three model parameters can be determined from the relay feedback response for true FOPDT processes. Therefore, the identification consists of the following steps:

(0) Record the time to the peak amplitude D, the peak amplitude a, and the period of oscillation P_u.
(1) Set the dead time D as the time to the peak value (Figure 4.5).
(2) Compute the time constant τ from Equation 4.8. Note that Equation 4.8 is an implicit equation for τ that requires an iterative solution. One can use the relationship between the ultimate frequency and τ given below to obtain a first guess for the value of τ:

$$\tau = \frac{\tan(\pi - D\omega_u)}{\omega_u} \tag{4.10}$$

(3) Compute K_p from Equation 4.9.

Similarly, for an unstable FOPDT system, all three model parameters can be obtained directly from relay feedback responses:

$$G(s) = \frac{K_p e^{-Ds}}{\tau s + 1} \tag{4.11}$$

First, the output responses can be represented analytically, as shown in Figure 4.6. From the first half period, the expression becomes

$$y(t) = K_p h(e^{t/\tau} - 1) \quad \text{for} \quad 0 < t < D \tag{4.12}$$

$$y(t) = K_p h(e^{t/\tau} - 1) - 2K_p h(e^{(t-D)/\tau} - 1) \quad \text{for} \quad D < t < P_u/2 \tag{4.13}$$

Substituting the two boundary conditions $y(D) = a$ and $y(P_u/2) = 0$ into Equations 4.12 and 4.13, the time constant τ and the steady state gain K_p can be computed directly:

$$\tau = \frac{P_u/2}{\ln\left[1/\left(2e^{-D/\tau} - 1\right)\right]} \tag{4.14}$$

$$K_p = \frac{a}{h}(e^{D/\tau} - 1) \tag{4.15}$$

The identification procedure for the unstable process is exactly the same as that of the stable one, except that the equations to compute τ and K_p are Equations 4.14 and 4.15 respectively. Equation 4.14 also reveals that the condition for the existence of limit cycle requires $D/\tau < \ln 2$.

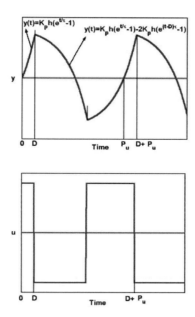

Figure 4.6. Analytical expressions of relay feed back response for unstable FOPDT systems (shifted version)

4.2.1.2 Category 1b: Approximated First-order Plus Dead Time

This category also has the FOPDT model structure. Similar to category 1a, stationary cycling develops in the first cycle, but the sharp edge around the peak amplitude is not quite as obvious as category 1a. The second- and third-order systems with $D/\tau = 10$ fall into this category. Since the true process is not exactly an FOPDT system, reading off the dead time directly from the response (*e.g.* Figure 4.5B) can be erroneous. Similar to Luynen's approach, we first define the time to reach the peak amplitude a as t_a and to reach half of the peak amplitude $a/2$ as $t_{a/2}$. Following the analytical expression in Equations 4.6 and 4.7, we have

$$a = K_p h \left(1 - e^{-t_a/\tau}\right) \tag{4.16}$$

$$\frac{a}{2} = K_p h \left(1 - e^{-t_{a/2}/\tau}\right) \tag{4.17}$$

Dividing Equation 4.16 by Equation 4.17, we can solve for τ using

$$2e^{-t_{a/2}/\tau} - e^{-t_a/\tau} = 1 \tag{4.18}$$

Once τ becomes available, we can solve for the other two model parameters, K_p and D, from ultimate properties:

$$D = \frac{\pi - \tan^{-1}(\tau\omega_u)}{\omega_u} \tag{4.19}$$

$$K_p = \frac{\sqrt{1+(\tau\omega_u)^2}}{K_u} \tag{4.20}$$

Therefore, the identification consists of the following steps:
(0) Record the time to the peak amplitude t_a, the time to one-half of the peak amplitude $t_{a/2}$, the peak amplitude a, and the period of oscillation P_u.
(1) Compute the time constant τ from Equation 4.18.
(2) Calculate the dead time from Equation 4.19.
(3) Calculate the steady state gain K_p from Equation 4.20.

This procedure enables us to find the approximate FOPDT model.

4.2.2 Identification of Category 2: Second-order Plus Small Dead Time

As pointed out earlier, if a small dead time to time constant ratio ($\varepsilon = D/\tau$) is specified, we have only two unknown parameters, as shown in Equation 4.3. After several numerical simulations, the default value of $\varepsilon = 0.001$ was found to work well for a wide range of parameter values. The time constant can be obtained from the phase angle information

$$-\pi = -\varepsilon\tau\omega_u - 2\tan^{-1}(\tau\omega_u) \tag{4.21}$$

and the steady state gain can be computed according to

$$K_p = \frac{1+(\tau\omega_u)^2}{K_u} \tag{4.22}$$

Therefore, the procedure consists of the following steps:
(0) Record the values of the peak amplitude a and the period of oscillation P_u.
(1) Compute the time constant τ from Equation 4.21.
(2) Calculate the steady state gain K_p from Equation 4.22.

This procedure enables us to find the SOPSDT model with the default setting of $\varepsilon = 0.001$. However, for certain cases, the model parameters obtained using the default value of $\varepsilon = 0.001$ may not be satisfactory. This is due to the fact that the ratio ε depends on the rate at which the oscillations are developed. Figure 4.7 presents the relay feedback responses obtained for SOPSDT processes with $\varepsilon = 0.001$, 0.005 and 0.01. Observations from Figure 4.7 indicate that, as ε increases, the normalized time constant (ratio of time constant τ_G defined by the peaks of the oscillations to the period of oscillation P_u) decreases. In other words, as D/τ increases it takes fewer cycles to reach the static oscillations. Figure 4.8 gives a plot showing the dependence of the D/τ ratio on the normalized time constant. A linear model is used to relate log(ε) to the normalized time constant τ_G / P_u:

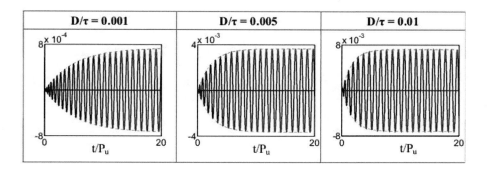

Figure 4.7. Relay feedback responses and global responses (defined by peaks) of SOPSDT processes with different D/τ ratios

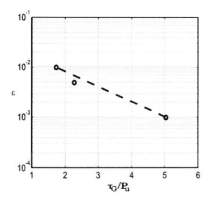

Figure 4.8. Relationship between the normalized global time constant (τ_G/P_u) and the dead time/time constant ratio ($\varepsilon = D/\tau$) for SOPSDT processes

$$\log(\varepsilon) = -0.3031\left(\frac{\tau_G}{P_u}\right) - 1.4767 \qquad (4.23)$$

Thus, with the value of τ_G / P_u from the relay feedback tests, we are able to calculate ε. Therefore, this more elaborate procedure consists of the following steps:
(0) Record the values of the peak amplitude a, the period of oscillation P_u and the time constant τ_G from the global response (Figure 4.7).
(1) Compute the value of the normalized time constant τ_G / P_u.
(2) Compute the value of ε using Equation 4.23.
(3) Compute the time constant τ from Equation 4.21.
(4) Calculate the steady state gain K_p from Equation 4.22.

This procedure enables us to find the ratio ε and the model parameters for the SOPSDT model.

4.2.3 Identification of Category 3: High Order

After several numerical simulations, the default value of $n = 5$ was found to work well for wide range of parameter values. Thus, if the order n is chosen for category 3 (e.g. $n = 5$ in Equation 4.4), we are left with two unknown parameters. They can be solved according to the ultimate properties via

$$\tau = \frac{\tan(\pi/n)}{\omega_u} \qquad (4.24)$$

$$K_p = \frac{\left(1 + (\tau \omega_u)^2\right)^{n/2}}{K_u} \qquad (4.25)$$

Similarly, the identification procedure becomes:
(0) Record the values of the peak amplitude a and the period of oscillation P_u.
(1) Compute the time constant τ from Equation 4.24 (with $n = 5$).
(2) Calculate the steady state gain K_p from Equation 4.25 (with $n = 5$).

This procedure enables us to find the parameters for the fifth-order process (HO model). However, for certain cases, the model parameters obtained using the default value of $n = 5$ might not be satisfactory. This is because the value of n depends on the rate at which the oscillations develop. Similar to the discussion in Section 3.3.2, we can relate the normalized time constant τ_G / P_u to the order n by Equation 4.26. A linear model is used to interpolate for n and τ_G / P_u between $n = 3$ and $n = 10$.

$$\log(n) = 1.9040 - 1.3736 \left(\frac{\tau_G}{P_u}\right) \qquad (4.26)$$

Thus, with the value of τ_G / P_u from the relay feedback tests, we are able to calculate the order n. Therefore, the more elaborate procedure consists of the following steps:
(0) Record the values of the peak amplitude a, the period of oscillation P_u, and the time constant of the curvature τ_G.
(1) Compute the value of the curvature factor $C = \tau_G / P_u$.
(2) Compute the value of n using Equation 4.26.
(3) Compute the time constant τ from Equation 4.24.
(4) Calculate the steady state gain K_p from Equation 4.25.

This procedure enables us to find the order as well as model parameters for the HO model.

4.2.4 Validation

To illustrate the appropriateness of the proposed classification, six typical examples representing the various categories are considered (Table 4.1). Relay feedback tests were conducted on all of these examples, and the relay feedback responses thus obtained are shown in Figure 4.9 under the heading "true". Time-domain responses clearly indicate that examples 1–3 can be classified as approximated FOPDT systems (category 1b), example 4 is an SOPSDT system (category 2), and example 5 and 6 can be classified as HO processes (category 3). The ultimate properties computed from the relay experiments are also presented in Table 4.1.

The equivalent models in the three categories for these six examples were formulated using the identification procedures described in the previous section. The 18 equivalent models thus obtained are presented in Table 4.2. Note that each equivalent model has the same values of K_u and ω_u but very different model structures. Relay experiments were conducted on all three equivalent models, as shown in Figure 4.9. The results show that the correct model structure reproduces the relay feedback response and that the mismatched model structures give completely different curve shapes, despite having the same K_u and P_u. The time-domain responses confirm that examples 1 to 3 belong to category 1, example 4 belongs to category 2, and examples 5 and 6 belong to category 3.

The models identified in Table 4.2 also reveal the importance of applying appropriate model structure. In example 1, using the same values of K_u and ω_u, the SOPSDT model gives a steady state gain 1600 times the true value, and in example 4 the FOPDT model structure results in an unstable system. The results clearly indicate the need to extract model structure information from relay feedback tests.

Table 4.1. Processes studied and corresponding ultimate properties

Example	True process	K_u	P_u
1	$\dfrac{e^{-10s}}{(s+1)^3}$	1.274	25.35
2	$\dfrac{e^{-15s}}{(s+1)^{15}}$	1.274	59.34
3	$\dfrac{e^{-10s}}{(s+1)^2}$	1.274	23.36
4	$\dfrac{e^{-0.05s}}{(25s+1)^2}$	862.9	5.434
5	$\dfrac{e^{-0.6s}}{(6s+1)^8}$	1.8424	91.94
6	$\dfrac{e^{-0.001s}}{(s+1)^{20}}$	1.342	39.33

60 Autotuning of PID Controllers

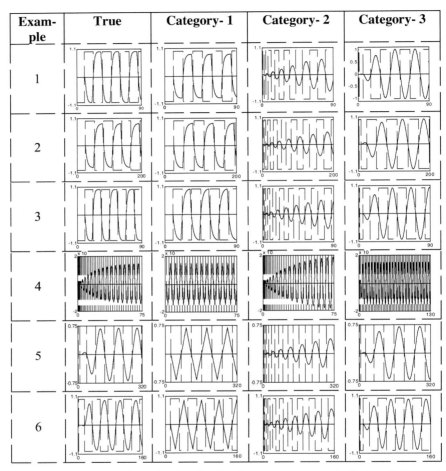

Figure 4.9. Reproduction of relay feedback responses (controlled variable, solid lines; manipulated variable, dashed lines) for six examples with the assumption of different model structures (from left to right: true process, FOPDT, SOPSDT, and HO)

Shape of Relay 61

Table 4.2. True process and equivalent models by assuming different model structures

Example	True process	FOPDT	SOPSDT	HO
1	$\dfrac{e^{-10s}}{(s+1)^3}$	$\dfrac{0.8724e^{-10.8567s}}{(1.9574s+1)}$	$\dfrac{1615.9378e^{-0.1830s}}{(182.95s+1)^2}$	$\dfrac{2.2648}{(2.9310s+1)^5}$
2	$\dfrac{e^{-15s}}{(s+1)^{15}}$	$\dfrac{0.8970e^{-24.9161s}}{(5.2159s+1)}$	$\dfrac{1615.9378e^{-0.1830s}}{(182.95s+1)^2}$	$\dfrac{2.2656}{(6.8613s+1)^5}$
3	$\dfrac{e^{-10s}}{(s+1)^2}$	$\dfrac{1.0169e^{-10.0017s}}{(2.4497s+1)}$	$\dfrac{1615.9645e^{-0.1686s}}{(168.6s+1)^2}$	$\dfrac{2.2658}{(2.7009s+1)^5}$
4	$\dfrac{e^{-0.05s}}{(25s+1)^2}$	$\dfrac{0.0234e^{-1.3233s}}{(21.2556s-1)}$	$\dfrac{2.3822e^{-0.0392s}}{(39.2s+1)^2}$	$\dfrac{0.0033}{(0.6283s+1)^5}$
5	$\dfrac{e^{-0.6s}}{(6s+1)^8}$	$\dfrac{12.0599e^{-23.6439s}}{(400.6714s+1)}$	$\dfrac{1118.8082e^{-0.6637s}}{(663.7s+1)^2}$	$\dfrac{1.5661}{(10.6313s+1)^5}$
6	$\dfrac{e^{-0.001s}}{(s+1)^{20}}$	$\dfrac{4.6861e^{-10.835s}}{(47.9111s+1)}$	$\dfrac{1533.106e^{-0.2839s}}{(283.91s+1)^2}$	$\dfrac{2.1494}{(4.5479s+1)^5}$

The above categorization can be validated in the frequency domain by evaluating the integrated absolute error (IAE) for each of the equivalent models. Multiplicative error is employed here.

$$e_{(\omega)} = \frac{G_m(j\omega) - G(j\omega)}{G(j\omega)} \qquad (4.27)$$

where G is the true process and G_m is the derived model. The IAE is evaluated between $0.1\omega_u$ and $10\omega_u$:

$$IAE_\omega = \int_{0.1\omega_u}^{10\omega_u} |e_{(\omega)}| d\omega \qquad (4.28)$$

The numerical values of IAE_ω are presented in Table 4.3. By comparing the values of IAE_ω of the models under categories 1, 2 and 3 of a particular true process, one can easily identify the category to which the true process belongs. The category to which the true process belongs offers the lowest IAE, thereby validating the proposed categorization. Even though all three model structures give small errors at the ultimate frequency, the correct category results in the lowest overall IAE_ω. With the help of IAE_ω values computed for different model structure processes over a wide range of D/τ values, as well as different orders, a quantitative classification can be made, as shown in Figure 4.4.

Table 4.3. Comparison of integrated frequency response errors by assuming different model structures

Example	True process	IAE		
		FOPDT	SOPSDT	HO
1	$\dfrac{e^{-10s}}{(s+1)^3}$	1.3083×10^{-4}	7.5365×10^{-3}	9.9724×10^{-4}
2	$\dfrac{e^{-15s}}{(s+1)^{15}}$	1.0958×10^{-4}	7.5468×10^{-3}	9.9986×10^{-4}
3	$\dfrac{e^{-10s}}{(s+1)^2}$	3.7721×10^{-5}	7.5233×10^{-3}	9.9337×10^{-4}
4	$\dfrac{e^{-0.05s}}{(25s+1)^2}$	9.9766×10^{-3}	2.3097×10^{-4}	5.1679×10^{-3}
5	$\dfrac{e^{-0.6s}}{(6s+1)^8}$	1.1144×10^{-3}	5.6718×10^{-3}	4.7446×10^{-4}
6	$\dfrac{e^{-0.001s}}{(s+1)^{20}}$	1.1952×10^{-3}	7.309×10^{-3}	9.3339×10^{-4}

4.3 Implications for Control

After identifying the appropriate model structure and associated model parameters, different tuning rules can be designed to achieve improved performance.

4.3.1 Proportional–Integral–Derivative Control

4.3.1.1 Category 1: First-order Plus Dead Time

Following Luyben's approach, different tuning formulas can be applied for different D/τ values. PI controllers are used here, but the approach can be extended to PID controllers with little difficulty.

1. $D/\tau < 0.1$

    For processes in category 1 having D/τ ratios less than 0.1, the Tyreus–Luyben tuning rule is found to be suitable.

    The Tyreus–Luyben tuning equations for PI controller are

$$K_c = \frac{K_u}{3.2} \quad (4.29)$$

$$\tau_I = 2.2 P_u \quad (4.30)$$

2. $0.1 \leq D/\tau \leq 1$

The minimum ITAE tuning rule developed by Rovira is found to be suitable for the FOPDT processes in category 1 with D/τ ratios ranging from 0.1 to 1.

The ITAE tuning equations for a PI controller are

$$K_c = \frac{0.586}{K_p}\left(\frac{\tau}{D}\right)^{0.916} \qquad (4.31)$$

$$\tau_I = \frac{\tau}{1.03 - 0.165\left(\dfrac{D}{\tau}\right)} \qquad (4.32)$$

3. $D/\tau > 1$

For processes in category 1 having D/τ ratios greater than 1, the PI controller with the IMC tuning rule is found to be suitable.

The IMC tuning equations for a PI controller are

$$\lambda = \max(1.7D, 0.2\tau) \qquad (4.33)$$

$$K_c = \frac{\tau + \dfrac{D}{2}}{K_p \lambda} \qquad (4.34)$$

$$\tau_I = \tau + \frac{D}{2} \qquad (4.35)$$

The various tuning rules for the FOPDT processes with different D/τ ratios are summarized in Table 4.4. For unstable FOPDT systems, the tuning rules given by Tan et al. [5], Huang and Chen [6], Marchetti et al. [7], and Jacob and Chidambaram [8] can be used.

Table 4.4 Tuning rules for FOPDT processes with different D/τ ratios

	$D/\tau < 0.1$	$0.1 \leq D/\tau \leq 1$	$D/\tau > 1$
Method	TL	ITAE	IMC
	$K_c = \dfrac{K_u}{3.2}$ $\tau_I = 2.2 P_u$	$K_c = \dfrac{0.586}{K_p}\left(\dfrac{\tau}{D}\right)^{0.916}$ $\tau_I = \dfrac{\tau}{1.03 - 0.165\left(\dfrac{D}{\tau}\right)}$	$\lambda = \max(1.7D, 0.2\tau)$ $K_c = \dfrac{\tau + \dfrac{D}{2}}{K_p \lambda}$ $\tau_I = \tau + \dfrac{D}{2}$

4.3.1.2 Category 2: Second-order Plus Small Dead Time

For SOPSDT systems, Ziegler–Nichols tuning gives poor performance. It should also be emphasized that this category is just an approximate modeling of the true process, so modeling error should be expected. For the SOPSDT processes with small values of D/τ (i.e. $D/\tau = 0.01 - 0.001$), we find the following rules appropriate. For a PI controller, first set

$$\tau_I = 2\tau \tag{4.36}$$

Then find that value of K_c that gives a 45° phase margin. For $D/\tau = 0.001$ this gives

$$K_c = \frac{2((0.0432\tau\omega_u)^3 + 0.0432\tau\omega_u)}{K_p(\sqrt{4(0.0432\tau\omega_u)^2 + 1})} \tag{4.37}$$

4.3.1.3 Category 3: High Order

Again, this is just an approximate model of a large variety of processes (Figure 4.4); therefore, a conservative tuning rule should be devised. For HO processes with orders ranging from 3 to 10, first, set the reset time to

$$\tau_I = (n-1)\tau \tag{4.38}$$

Next, adjusted K_c to give a maximum closed-loop log modulus L_c^{max} of 3 dB. For the default value of $n = 5$, this gives a very simple expression for K_c:

$$K_c = \frac{1}{K_p} \tag{4.39}$$

4.3.2 Results

After the exact category of the true process and the appropriate control strategy had been identified, closed-loop studies were carried out on all six examples listed in Table 4.1.

PI controllers were designed on the basis of the identified models. For example, for the third-order plus dead time process of example 1, three different PI controllers were designed according to the FOPDT, SOPSDT, and HO model structures (e.g. first row of Table 4.2) using the tuning rules presented in Table 4.4, and Equations 4.36–4.39. This procedure was repeated for all six examples. Table 4.5 gives the controller settings for the examples studied. The effects of model structures on closed-loop performance can thus be compared. Closed-loop studies were carried out on the true processes. The SP responses of three different controller settings (from different model structures) on the six examples are presented in Figure 4.10. A close look at the responses reveals the following.

Table 4.5. PI controller parameters for different examples

True process	FOPDT-IMC		SOPSDT-PM		HO- L_c^{max}	
	K_c	τ_I	K_c	τ_I	K_c	τ_I
$\dfrac{e^{-10s}}{(s+1)^3}$	0.4587	7.3858	0.0023	365.9	0.4418	11.724
$\dfrac{e^{-15s}}{(s+1)^{15}}$	0.4652	17.6739	0.0023	856.64	0.4416	27.4452
$\dfrac{e^{-10s}}{(s+1)^2}$	0.4309	7.4506	0.0023	337.2	0.4415	10.8036
	FOPDT-TL					
$\dfrac{e^{-0.05s}}{(25s+1)^2}$	269.6448	11.9548	1.5279	78.4	303.15	2.5132
$\dfrac{e^{-0.6s}}{(6s+1)^8}$	0.5758	202.268	0.0033	1327.4	0.6387	42.5252
	FOPDT-ITAE					
$\dfrac{e^{-0.001s}}{(s+1)^{20}}$	0.4880	48.2641	0.0024	567.82	0.4654	18.1916

For examples 1–3, the responses obtained from the controller designed using the FOPDT model structure are superior to the other responses. This is because examples 1–3 fall into category 1, for which a PI controller with IMC tuning is the recommended controller ($D/\tau > 1$). The controller settings obtained using the SOPSDT model structure give very slow responses, whereas those obtained by assuming an HO process result in undershoot responses.

For example 4, the correct model structure (SOPSDT) gives a reasonable SP response when compared with those of the other two model structures. The TL tuning (a conservative Ziegler–Nichols type of tuning) produces unstable responses for the second-order system with small dead-time-to-time-constant ratio, and the assumption of an HO model structure also fails to maintain stability. The reason is that this is almost a double integrator process, which is difficult to control.

For examples 5 and 6, the responses obtained from the controller designed using the equivalent HO models are superior to the other responses. The FOPDT model structure results in undershoot responses, whereas the SOPSDT equivalent model gives even more sluggish SP responses. Note that the controller gain K_c of the SOPSDT model is almost two orders of magnitude smaller than that of the reasonable model.

A special case under the SOPSDT category for which the default value of $\varepsilon = 0.001$ does not work very well is also considered below. Consider the SOPSDT process represented by

66 Autotuning of PID Controllers

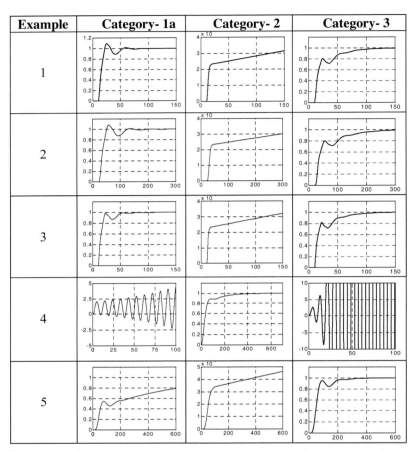

Figure 4.10. Comparison of SP responses of six examples with PI controller

$$G = \frac{e^{-0.01s}}{(2s+1)^2} \tag{4.40}$$

The relay feedback test output response indicates that the process belongs to category 2, *i.e.* an SOPSDT process with a model structure given by Equation 4.3. The model parameters obtained using the procedure described in Section 3.2 with the default value of $\varepsilon = 0.001$ is given by

$$G_m = \frac{6.027 e^{-0.05s}}{(5s+1)^2} \tag{4.41}$$

The controller settings obtained ($K_c = 0.6038$ and $\tau_I = 10$) using Equations 4.36 and 4.37 with the above model parameters result in a slow SP response (Figure 4.11). Hence, there is a need to select an appropriate value of ε using the procedure described in Section 4.2. The model parameters thus obtained with the modified value of $\varepsilon = 0.00685$ are given by the equation

$$G_m = \frac{0.8709e^{-0.013s}}{(1.8978s+1)^2} \tag{4.42}$$

The controller settings obtained ($K_c = 4.1794$ and $\tau_I = 3.7956$) using Equations 4.36 and 4.37 with the above model parameters result in a better SP response (Figure 4.11), on par with that obtained using the true process ($K_c = 3.6398$ and $\tau_I = 4$). Thus, it is clear that the appropriate model structure with suitable tuning rules offers better closed-loop performance. More importantly, the improvement is achieved by taking the shape factor in the relay feedback response into account.

In a real process environment, measurement noise is unavoidable. The proposed method was tested against measurement noise. In the context of system identification, the noise-to-signal ratio (NSR) can be expressed as

$$NSR = \frac{mean[abs(noise)]}{mean[abs(signal)]} \tag{4.43}$$

where abs(.) denotes the absolute value and mean (.) represents the mean value.

The following two FOPDT processes are used to illustrate the effect of process noises:

$$G(s) = \frac{e^{-0.1s}}{(s+1)} \tag{4.44}$$

$$G(s) = \frac{e^{-10s}}{(s+1)} \tag{4.45}$$

In the case of the process represented by Equation 4.44, relay feedback tests were performed with $NSR = 0$ and 1/5 and with a relay height of 1. The relay feedback responses thus obtained are shown in Figure 4.12A. The limit cycle data were computed by taking the average of the fictitious peaks around the peak. Two cycles were employed to compute the average values of the limit cycle data. The FOPDT

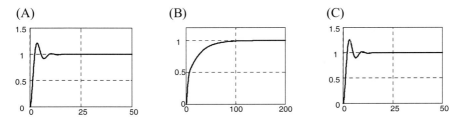

Figure 4.11. Comparison of closed-loop responses of an SOPSDT process (special case) for SP tracking: (A) controller settings obtained using the true process, (B) controller settings obtained using the (SOPSDT) model with the default value of ε = 0.001, and (C) controller settings obtained using the SOPSDT model with ε obtained from Equation 4.23 ($\varepsilon = 0.00685$)

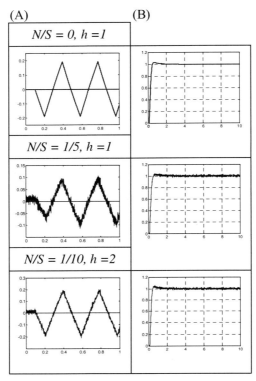

Figure 4.12. Performance of an FOPDT process (D/τ = 0.1) with different levels of measurement noise (*NSR* = 0, $1/5$, and $1/10$): (A) relay feedback responses, (B) SP responses with a PI controller

equivalent model (Table 4.6) was formulated using the procedure described in Section 4.1.1. Tyreus–Luyben tuning rules (Equations 4.29 and 4.30) were used to tune the PI controller settings (Table 4.6). The SP responses thus obtained are shown in Figure 4.12B and are on a par with the responses obtained for the noise-free process. However, when the process was corrupted with noise there was model mismatch in the FOPDT equivalent model derived. This mismatch can be alleviated by increasing the strength of the signal ($h=2$, $NSR=1/10$). Better FOPDT equivalent model parameters and controller settings were obtained (Table 4.6) by following a procedure similar to that described above. Figure 4.12A shows that the closed-loop performance results in satisfactory SP responses even under the influence of measurement noise.

Similarly, for the process represented by Equation 4.45, relay feedback tests were performed with $NSR=0$ and $1/10$ and with a relay height of 1. The corresponding relay feedback responses thus obtained are shown in Figure 4.13A. The limit cycle data were computed in a manner similar to that carried out for the process represented by Equation 4.44. The FOPDT equivalent model was formulated

Shape of Relay   69

Table 4.6. True process, equivalent models and corresponding PI controller settings by assuming different NSR (FOPDT with $D/\tau = 0.1$ and $D/\tau = 10$)

		True	N/S = 0	N/S = 1/10	N/S = 1/5
$D/\tau = 0.1$		$\dfrac{e^{-0.1s}}{(s+1)}$	$\dfrac{1.0091e^{-0.1s}}{(1.0118s+1)}$	$\dfrac{0.5526e^{-0.1036s}}{(0.5414s+1)}$	$\dfrac{0.3962e^{-0.1089s}}{(0.3717s+1)}$
	K_c	4.1900	4.1900	4.383	3.9527
	τ_I	0.8404	0.8404	0.8382	0.8492
		True	N/S = 0	N/S = 1/20	N/S = 1/10
$D/\tau = 10$		$\dfrac{e^{-10s}}{(s+1)}$	$\dfrac{0.9998e^{-10.0010s}}{(1.0258s+1)}$	$\dfrac{1.0026e^{-9.1667s}}{(1.6449s+1)}$	$\dfrac{1.2527e^{-8.7135s}}{(2.3101s+1)}$
	K_c	0.3529	0.3525	0.3986	0.3766
	τ_I	6	6.0263	6.2282	6.6668

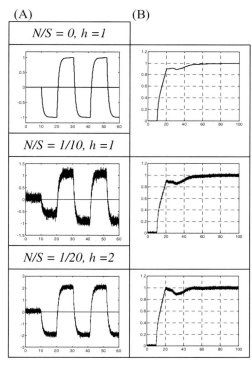

Figure 4.13. Performance of FOPDT process ($D/\tau = 10$) with different levels of measurement noise (*NSR* = 0, $1/5$, and $1/10$): (A) relay feedback responses, (B) SP responses with a PI controller

using the procedure described in Section 4.1.2. PI controller settings were obtained using the tuning rules given in Equajtion 4.33–4.35. Figure 4.13B shows the SP responses obtained using the above controller settings. Even though the closed-loop response obtained for $NSR = 1/10$ is on a par with that obtained for the noise-free process, the FOPDT model formulated has mismatch in the estimated model parameters (Table 4.6). An increase in the strength of the signal ($NSR = 1/20$, $h = 2$) can alleviate the model parameter mismatch. The PI controller settings derived using the above model parameters result in a satisfactory SP response, as shown in Figure 4.13B.

The observations from Figures 4.12B and 4.13B indicate that the proposed method works well even in the presence of measurement noise, resulting in satisfactory closed-loop performance.

4.3.3 Extension

Because the model structure and corresponding parameters are available, dead-time compensation or higher order compensation can be provided whenever necessary. For the purpose of illustration, IMC was used to design the controllers for an FOPDT system with a large D/τ and for HO systems.

4.3.3.1 Dead-time-Dominant Process

For processes in category 1b, a dead-time compensator can be designed to improve the performance. Given the model G_m and IMC filter F (a first-order filter)

$$G_m = \frac{K_p e^{-Ds}}{\tau s + 1} \tag{4.46}$$

$$F = \frac{1}{\lambda s + 1} \tag{4.47}$$

the conventional controller K using the IMC design becomes

$$K = \frac{G_c}{1 - G_c G_m} = \frac{1}{K_p} \frac{(\tau s + 1)}{\left[(\lambda s + 1) - e^{-Ds}\right]} \tag{4.48}$$

In this work, the filter time constant is set to

$$\lambda = 2\tau \tag{4.49}$$

Example 1 ($D/\tau = 5.5465$) is used to illustrate the potential improvement; Figure 4.14 shows that better SP responses can be obtained for systems with large D/τ values.

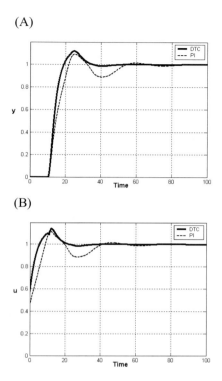

Figure 4.14. SP responses for a process with large dead time (category 1b) using a dead-time compensator (solid), and a PI controller (dashed lines)

4.3.3.2 Higher Order Process

For processes in category 3, a higher order controller can be used to improve the control performance. Again, IMC control was employed with the following model G_m and filter F:

$$G_m = \frac{K_p}{(\tau s + 1)^5} \qquad (4.50)$$

$$F = \frac{1}{(\lambda s + 1)^5} \qquad (4.51)$$

The recommended value for λ is 0.45τ. Thus, with the known values of G_m and F, the equivalent controller in the conventional feedback structure becomes

$$K = \frac{1}{5K_p \lambda s}\left[\frac{(\tau s+1)^5}{(\lambda^4/5)s^4 + \lambda^3 s^3 + 2\lambda^2 s^2 + 2\lambda s + 1}\right] \qquad (4.52)$$

72   Autotuning of PID Controllers

Example 5 is used to illustrate the potential improvement. Figure 4.15 shows that a better load response is achieved using the higher order controller without exciting the manipulated variable excessively.

4.4 Conclusion

The shapes of relay feedback responses are useful in extracting additional information about process dynamics. From a methodical analysis of the shape information, different processes can be broadly classified into three major categories. Analytical expressions for the responses of stable and unstable FOPDT processes can be used to derive all three model parameters. From the insight gained, the identification procedures for different processes under various categories were evolved. Different tuning rules were employed to find appropriate PID controller settings. Procedures are tested against linear systems with and without noise. Further, dead-time

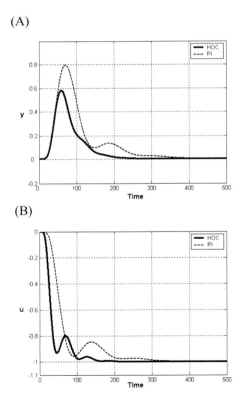

Figure 4.15. Load responses for a higher order process (category 3) using a higher order controller (solid lines) and a PI controller (dashed lines)

compensation (for category 1b) and higher order compensation (for category 3) can also be devised whenever necessary. The results show that the proposed method results in improved autotuning in a straightforward manner. Thus, shape information is useful in inferring the correct model structure of the process and also in selecting the appropriate control strategy to offer improved performance without the need for any additional testing.

4.5 References

1. Luyben WL. Getting more information from relay feedback tests. Ind. Eng. Chem. Res. 2001;40(20):4391.
2. Åström KJ, Hägglund T. Automatic tuning of simple regulators with specifications on phase and amplitude margins. Automatica 1984;20:645.
3. Friman M, Waller KV. Autotuning of multiloop control systems. Ind. Eng. Chem. Res. 1994;33(7):1708.
4. Wang QG, Hang CC, Zou B. Low-order modeling from relay feedback. Ind. Eng. Chem. Res. 1997;36(2):375.
5. Tan KK, Wang QG, Lee TH. Finite spectrum assignment control of unstable time delay processes with relay tuning. Ind. Eng. Chem. Res. 1998;37(4):1351.
6. Huang HP, Chen CC. Autotuning of PID controllers for second order unstable process having dead time. J. Chem. Eng. Jp. 1999;32(4):486.
7. Marchetti G, Scali C, Lewin DR. Identification and control of open-loop unstable processes by relay methods. Automatica 2001;37:2049.
8. Jacob EF, Chidambaram M. Design of controllers for unstable first-order plus time delay systems. Comput. Chem. Eng. 1996;20(5):579.

5
Improved Relay Feedback

Luyben [1] pioneered the use of relay feedback tests for system identification. The ultimate gain and ultimate frequency from the relay feedback test are used to fit a typical transfer function (*e.g.* first-, second- or third-order plus dead time system). As mentioned in Chapter 3, it can lead to significant errors in the ultimate gain and ultimate frequency approximation (*e.g.* 5–20% error in K_u) for typical transfer functions in a control system. The errors come from the linear approximation (describing function analysis) to a nonlinear element. The square type of output from the relay is approximated with the principal harmonic from the Fourier transform [2,3] and the ultimate gain is computed accordingly. Several attempts have been proposed to overcome this inaccuracy. Li *et al.* [4] use two relay tests to improve the estimation of K_u and ω_u. Chang *et al.* [3] employ the concept of a discrete-time system to give a better estimation of ω_u. Notice that, in these attempts, an ideal relay is employed in the experiments and modifications are made *afterward*. Since, the source of the error comes from sine-wave approximation of a square type of oscillation, a straightforward approach to overcome the inaccuracy is to re-design the experiment (instead of taking remedial action afterward). That is, to produce a more sine-wave-like output using a different type of relay.

In this chapter we are trying to design an experiment such that better accuracy can be achieved in estimating ultimate gain and ultimate frequency. Section 5.1[1] provides the fundamentals for the saturation relay and an in-depth analysis is given. Section 5.2 discusses the experimental design and a procedure is given. Single-loop and multivariable examples are illustrated in Section 5.3.

---

[1] Readers who are interested in the method itself can skip this section and go directly to Section 5.2.

5.1 Analysis

5.1.1 Ideal (On–Off) Relay Feedback

Autotuning based on relay feedback can be analyzed via a block diagram. Consider a feedback system (Figure 5.1) where $G(s)$ is a linear transfer function and N is a nonlinear element. If the input signal $e(t)$ to the nonlinear element is a sinusoidal wave

$$e(t) = a \sin \omega t \tag{5.1}$$

where a is the magnitude of the sinusoidal wave, then the output signal $u(t)$ of the nonlinear element is a square wave (Figure 5.2).

Since most control system analyses are based on linear theory, Fourier transformation is useful in this regard. The output of the nonlinear element can be expressed as

$$u(t) = A_0 + \sum_{n=1}^{\infty} A_n \cos n\omega t + B_n \sin n\omega t \tag{5.2}$$

where

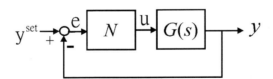

Figure 5.1. Nonlinear feedback system

(A) Ideal Relay

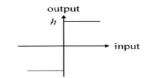

(B) Input-Output responses

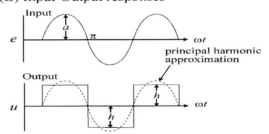

Figure 5.2. Input–output relationship for ideal (on–off)

$$A_0 = \frac{1}{2\pi} \int_0^{2\pi} u(t)\, d(\omega t) \tag{5.3}$$

$$A_n = \frac{1}{\pi} \int_0^{2\pi} u(t) \cos n\omega t\, d(\omega t) \tag{5.4}$$

$$B_n = \frac{1}{\pi} \int_0^{2\pi} u(t) \sin n\omega t\, d(\omega t) \tag{5.5}$$

Because the output $u(t)$ is an odd-symmetric function (*i.e.* $N(a)$ is unbiased and symmetric to *the origin*), the coefficients A_0 and A_n are equal to zero (*i.e.* $A_0 = 0$ and $A_n = 0$, $\forall n$). Therefore, Equation 5.2 becomes

$$u(t) = \sum_{n=1}^{\infty} B_n \sin n\omega t \tag{5.6}$$

Furthermore, if an ideal relay is employed (Figure 5.2), then the coefficients B_n become

$$B_n = \begin{cases} \dfrac{1}{n} \dfrac{4h}{\pi}, & n = 1,\ 3,\ 5,\ \cdots \\ 0, & n = 2,\ 4,\ 6,\ \cdots \end{cases} \tag{5.7}$$

The describing function analysis provides a tool for frequency-domain analysis for this nonlinear system. Only the principal harmonic is employed for the linear equivalence. That is, only the first Fourier coefficient is used for frequency-domain analysis. Therefore, the describing function becomes

$$N(a) = \frac{B_1 + jA_1}{a} \tag{5.8}$$

For the ideal relay, since $A_1 = 0$ and $B_1 = 4h/\pi$, we have

$$N(a) = \frac{4h}{\pi a} \tag{5.9}$$

Since a sustained oscillation is generated from a relay feedback test (*e.g.* Figure 5.2), the frequency of oscillation corresponds to the limit of stability, *i.e.*

$$1 + G(j\omega_u) N(a) = 0 \tag{5.10}$$

or the ultimate gain K_u becomes

$$\begin{aligned} K_u &= -\frac{1}{G(j\omega_u)} \\ &= N(a) \\ &= \frac{4h}{\pi a} \end{aligned} \tag{5.11}$$

Part of the success of the autotune identification comes from the fact that K_u and ω_u can be read directly from the experimental results (*e.g.* Figure 5.2B).

The results of Equation 5.11 clearly indicate that the ultimate gain K_u is estimated from the amplitude ratio of two sinusoidal waves at a given frequency ω_u (*i.e.* $u(t) = (4h/\pi)\sin\omega t$ over $e(t) = a\sin\omega t$). Obviously, the output of the relay is a square wave instead of a sinusoidal wave. This leads to an erroneous result in the estimated ultimate gain. Figure 5.2 shows the input–output relationship for an ideal relay. Here, the principal harmonic is used to approximate a square wave (Figure 5.2B). Chang *et al.* [3] point out that the truncation of the higher order terms (*i.e.* $n = 3, 5, 7, \cdots$) affects the ultimate gain and ultimate frequency estimation. Mathematically, it is difficult to include the high order terms in a linear analysis. Instead of including higher order terms, a straightforward approach is to redesign the relay feedback experiment. In other words, it is helpful to devise an experiment such that the output response of the relay is of more sine-wave-like, *i.e.* less square-wave-like.

5.1.2 Saturation Relay Feedback

Since the square-wave output (*e.g.* Figure 5.2B) comes from an abrupt slope change at the zero point (*i.e.* $e(t) = 0$ in Figure 5.2), the saturation relay provides a smooth transition around the zero point, as shown in Figure 5.3A. The saturation relay is characterized by two parameters: a relay height h and a slope k (Figure 5.3A). Therefore, the input of the relay is limited by a maximum $\bar{a}$, where

$$\bar{a} = \frac{h}{k} \tag{5.12}$$

That is, if the input to the relay is less than $\bar{a}$ ($|e| \leq \bar{a}$), then the output is proportional to the input with a factor k

$$u = ke \tag{5.13}$$

However, if the input to the relay is greater than $\bar{a}$ ($|e| > \bar{a}$)), then the output of the relay is limited by h

$$u = h \tag{5.14}$$

or

$$u = -h \tag{5.15}$$

With the saturation relay inserted in the feedback loop, the output of the relay shows a *less* square-wave-like response, *e.g.* a sine wave with an upper (or a lower) limit. The height of the output response is limited by h ($h = k\bar{a}$). The output of the saturation relay can be characterized analytically. Consider a saturation relay feedback system. The input to the nonlinear element is a sinusoidal wave with an amplitude a (Figure 5.3B), *i.e.*

(A) Saturation Relay

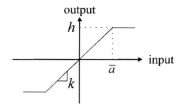

(B) Input-Output responses

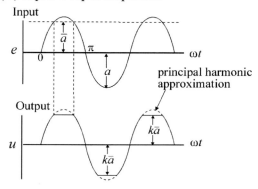

Figure 5.3. Input–output relationship for saturation relay

$$e(t) = a \sin \omega t \qquad (5.16)$$

The output to the nonlinear element $u(t)$ looks like a truncated sinusoidal wave and the closeness of this output response to a sine wave depends a great deal on the slope k chosen. The angle γ (Figure 5.4) gives a simple measure to characterize the relay output.

$$\gamma = \sin^{-1}\left(\frac{\overline{a}}{a}\right) \qquad (5.17)$$

Since the relay output is a periodic function, considering a half period, if the phases lie between γ and $\pi - \gamma$, then the output is equal to h and the sine-wave-like responses remain for $\omega t < \gamma$ and $\omega t > \pi - \gamma$, as shown in Figure 5.4. Obviously, the value γ depends on the slope k. If $k \to \infty$, then we have

$$\gamma = \lim_{k \to \infty} \sin^{-1}\left(\frac{(h/k)}{a}\right) = 0 \qquad (5.18)$$

Then, the output becomes a square wave. With this measure, the relay output can be expressed as

80   Autotuning of PID Controllers

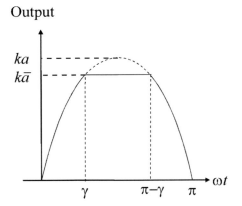

Figure 5.4. Graphical interpretation of the angle γ

$$u(t) = \begin{cases} \dfrac{h}{\sin\gamma}\sin\omega t & 0 \le \omega t < \gamma \text{ and } \pi - \gamma < \omega t \le \pi \\ h & \gamma \le \omega t \le \pi - \gamma \end{cases} \quad (5.19)$$

Since the principal harmonic is employed for the linear approximation, the Fourier transformation of $u(t)$ is useful for the purpose of analysis:

$$u(t) = \sum_{n=1}^{\infty} B_n \sin n\omega t \quad (5.20)$$

where

$$B_n = \frac{2}{\pi}\int_0^\pi u(t)\sin n\omega t \, d(\omega t) \quad (5.21)$$

Since the term γ plays an important role in the frequency-domain analysis, the relationship between γ and frequency responses is studied. Consider the following cases.

(A) $0 < \gamma < \pi/2$ ($\infty > k > h/a$). For this general case, substituting $u(t)$ (Equation 5.19) into Equation 5.21, we have

$$B_n = \begin{cases} \dfrac{2h}{\pi}\left[\dfrac{1}{\sin\gamma}\left(\dfrac{\sin(1-n)\gamma}{1-n} - \dfrac{\sin(1+n)\gamma}{1+n}\right) - \dfrac{1}{n}\left(\cos n(\pi-\gamma) - \cos\gamma\right)\right] & , n = 1, 3, 5, \cdots \\ 0 & , n = 2, 4, 6, \cdots \end{cases} \quad (5.22)$$

The expression for the odd coefficients (*i.e.* $B_1, B_3, B_5, \cdots$) differs from that of an ideal relay [3] and the even coefficients remain zero. Thus, the describing function becomes

$$N(a) = \frac{2h}{\pi \bar{a}} \left[\left(\sin^{-1} \frac{\bar{a}}{a} \right) + \left(\frac{\bar{a}}{a} \sqrt{1 - \left(\frac{\bar{a}}{a} \right)^2} \right) \right] \qquad (5.23)$$

where $\bar{a} = h/k$. Since the higher order terms (Equation 5.22) are neglected, Figure 5.5 clearly shows that the principal harmonic approximation cannot exactly describe the output response (*e.g.* Figure 5.5B).

(B) $\gamma = 0$ ($k \to \infty$). Let us first consider an asymptotic case when the slope of the saturation relay approaches infinity, *i.e.* $k \to \infty$ (Figure 5.3A). In this case, $\bar{a}$ becomes zero (Equation 5.12), γ is zero (Equation 5.17) and, subsequently, the saturation relay is reduced to an ideal relay (Figure 5.2). The coefficients B_n of the Fourier expansion can be derived from Equation 5.22. After some algebraic manipulation, we have

$$B_n = \lim_{\gamma \to 0} \frac{2h}{\pi} \left[\frac{1}{\sin \gamma} \left(\frac{\sin(1-n)\gamma}{1-n} - \frac{\sin(1+n)\gamma}{1+n} \right) \right.$$

$$\left. - \frac{1}{n} \left(\cos(\pi - \gamma) - \cos \gamma \right) \right] \qquad (5.24)$$

$$= \frac{1}{n} \frac{4h}{\pi} \quad , n = 1, 3, 5, \cdots$$

and

$$B_n = 0 \quad , n = 2, 4, 6, \cdots \qquad (5.25)$$

Since the principal harmonic B_1 is employed in the describing function analysis, $N(a)$ becomes

$$N(a) = \lim_{\bar{a} \to 0} \frac{2h}{\pi} \left(\frac{1}{\bar{a}} \sin^{-1} \frac{\bar{a}}{a} + \frac{\sqrt{a^2 - \bar{a}^2}}{a^2} \right)$$

$$= \frac{4h}{\pi a} \qquad (5.26)$$

Again, the principal harmonic approximation cannot describe the true output response. Figure 5.5A compares the true output response (solid line) with the principal harmonic approximation (dashed line).

(C) $\gamma = \pi/2$ ($k = h/a$). Let us consider another asymptotic case. That is, the slope is carefully chosen such that $\bar{a} = a$ (or $k = h/a$). In this case, the output of the relay is exactly a sine wave (*e.g.* Figure 5.5C). Therefore, the

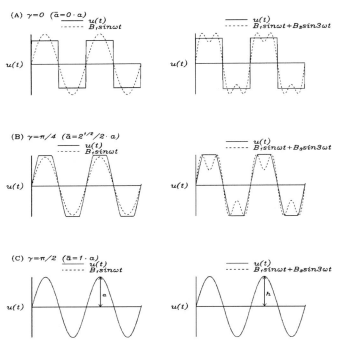

Figure 5.5. The wave shape of the output to the nonlinear element for different slopes (solid: true output; dashed: approximations)

Fourier coefficients can be found by substituting $\gamma = \pi/2$ into Equation 5.22. Here, we have

$$B_n = \begin{cases} h & n = 1 \\ 0 & \text{otherwise} \end{cases} \quad (5.27)$$

In this case, only the primary harmonic term exists ($B_1 = h$ and $B_n = 0$ for $n \geq 2$) and it gives the exact solution. Thus, the output of the saturation relay is

$$u(t) = h \sin \omega t \quad (5.28)$$

Equation 5.28 shows that the saturation relay gives a pure sinusoidal wave and that the output lags behind the input by $-180°$. Obviously, this is exactly the conventional sine-wave test and, from the definition, the ultimate gain K_u is

$$K_u = \frac{h}{a} \quad (5.29)$$

From the describing function analysis, $N(a)$ can be found by substituting $\gamma = \pi/2$ into Equation 5.23:

$$N(a) = \frac{h}{a} \tag{5.30}$$

Comparing Equation 5.30 with Equation 5.29, it is clear that *no* approximation is involved in this estimation.

The analyses of these three cases show that the saturation relay is a generalization of the ideal relay. More importantly, better estimates of K_u and ω_u can be achieved by adjusting the slope of the relay. For example, when $\gamma = 0$ we have an ideal relay, and as γ increases to $\pi/2$ the experiment becomes a conventional sine-wave test. Therefore, it provides the flexibility in finding more accurate values of K_u and ω_u.

5.1.3 Potential Problem

The improvement in the estimates of K_u and ω_u does not come without any potential problem. One possible case is that if the slope chosen is too small (or a is smaller than $\bar{a}$ or h/k), then a limit cycle may not exist. This can be analyzed from frequency responses. Notice that the condition for the existence of a sustained oscillation is

$$1 + G(j\omega_u)N(a) = 0 \tag{5.31}$$

or

$$G(j\omega_u) = -\frac{1}{N(a)} \tag{5.32}$$

Equation 5.32 can be solved by plotting and the intersection corresponds to the crossover point (K_u and ω_u). For an ideal relay, the $-1/N$ loci starts from the origin and goes toward $-\infty$ (e.g. starting from the point "a" toward left in Figure 5.6). In terms of a saturation relay, the starting point of the $-1/N$ loci corresponds to $-1/k$ (Figure 5.6). As we decreases the slope, the starting point moves gradually to the left. If the starting point moves to the point "b", we still have an intersection and a limit will exist. However, if the slope is decreased further, such that the starting point moves over the point "c", then $-1/N$ does not intersect $G(j\omega)$ any longer and we do not have a limit cycle. Therefore, there exists a critical slope k_{min}; when the slope of the saturation relay is smaller than this value, the feedback system cannot generate a sustained oscillation. On the other hand, if the slope is chosen too large, then the relay approaches an ideal relay and the improvement in the estimate of K_u and ω_u disappears. Therefore, a trade-off has to be made in the selection of the slope. Furthermore, this critical slope k_{min} is related to $|G(j\omega)|$. That is:

$$k_{min} = \frac{1}{|G(j\omega_u)|} \tag{5.33}$$

84 Autotuning of PID Controllers

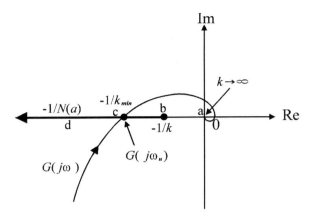

Figure 5.6. Loci of $G(j\tau)$ and $-1/N(a)$

The following example illustrates the trade-off.

Example 5.1 WB column

$$G(s) = \frac{x_D}{R} = \frac{12.8\,e^{-s}}{16.8s+1}$$

This FOPDT system has the following ultimate properties: $K_u = 2.1$ and $\omega_u = 1.608$. If an ideal relay is used ($k \to \infty$), then the test gives $\hat{K}_u = 1.71$ and $\hat{\omega}_u = 1.615$. This test shows an almost −20% error in the estimate of K_u. Furthermore, neither the input x_D nor the output R of the relay shows sine-wave-like responses (Figure 5.7A). If the slope k decreases to 5, then the system responses behave more sine-wave-like (Figure 5.7B) and the estimate of K_u becomes $\hat{K}_u = 1.94$ (8% error). Obviously, an improvement in the estimate of K_u can be seen using the saturation relay. If the slope decreases further to the critical slope ($k = k_{min} = 2.1$), then the input and output of the relay look exactly like a sine-wave (Figure 5.7C) and the estimates become $\hat{K}_u = 2.098$ and $\hat{\omega}_u = 1.607$. These are almost the exact values for K_u and ω_u. However, if the slope is chosen to be less than k_{min} (e.g. $k = 1.5$), then the relay fails to generate a sustained oscillation.∎

The above example clearly indicates that the saturation relay can improve the estimation of ultimate gain and ultimate frequency. However, attention has to be paid in the selection of the slope k.

5.2 Improved Experimental Design

5.2.1 Selection of the Slope of Saturation Relay

As mentioned earlier, a critical slope k_{min} exists to indicate the success/failure of a relay feedback test. Furthermore, this critical slope is system dependent (Equation 5.29). Qualitatively, we also understand that the smaller the slope k, the more accurate the estimate of K_u and ω_u which can be achieved given the assumption that

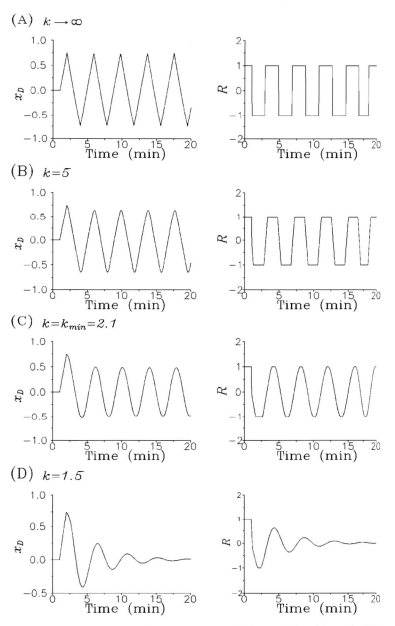

Figure 5.7. Relay feedback test for Example 5.1: (A) ideal relay ($k \to \infty$), (B) saturation relay ($k = 5$), (C) saturation relay ($k = k_{min} = 2.1$), (D) saturation relay ($k = 1.5 < k_{min}$)

the test is successful. However, in a relay feedback test a *quantitative* value of the slope should be given.

In order to determine the slope, a typical process transfer function for chemical processes is used to illustrate the trade-off between the success of an experiment and the accuracy of the estimate. Consider a transfer function of the form

$$G(s) = \frac{e^{-Ds}}{\tau s + 1} \qquad (5.34)$$

where D is the dead time and τ is the time constant. A range of D/τ is studied for different values of dimensionless slope k/k_{min} and, subsequently, percentage errors in K_u and ω_u are evaluated. Results (Figure 5.8) show that the improvement in the estimate levels off as $k \to 10 k_{min}$. Furthermore, the error in K_u ranges from −10% to −20% for these FOPDT systems with an ideal relay ($k \to \infty$) and the experiments tend to *underestimate* K_u. Several things become apparent immediately. First, generally, the slope should be less than $10 k_{min}$ in order to improve the estimates. Second, it is preferable to choose a slope of at least $1.4 k_{min}$ to avoid an unsuccessful relay feedback test ($k_{min} = K_u = 1/|G(j\omega_u)|$). Therefore, a simple rule of thumb is to select the slope as "1.4" times k_{min}. Notice that this is a safety factor for a class of transfer functions over a range of parameter values; for a given system, the true safety factor is actually system dependent,

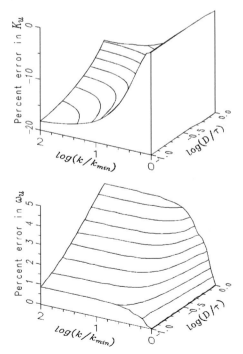

Figure 5.8. Percentage error in K_u and τ_u for an FOPDT system ($e^{-Ds}/(\alpha s + 1)$ with different values of D/ω

as shown in Figure 5.8. In order to test the *validity* of this method, let us consider a second-order example.

Example 5.2 High-purity distillation column [5]

$$G(s) = \frac{37.7e^{-10s}}{(7200s+1)(2s+1)}$$

For this system, the exact values for K_u and ω_u are 26.24 and 0.1315 respectively. For an ideal relay ($k \to \infty$), the ultimate gain found is 23.15. This corresponds to -11.7% error in K_u. As we decrease the slope to k_{min}, the almost exact value of K_u can be found ($K_u = 26.04$) (Figure 5.9). Furthermore, for $k = 1.4 k_{min}$ we can generate a sustained oscillation with an improved estimate in both K_u and ω_u (Figure 5.9). ∎

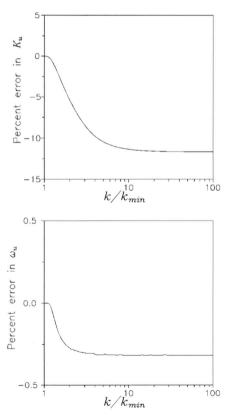

Figure 5.9. Percentage error in K_u and τ_u for Example 5.2

Example 5.3 Second-order system with RHP zero

$$G(s) = \frac{(-10s+1)e^{-s}}{(2s+1)(4s+1)}$$

This is a system with an inverse response. The exact values for K_u and ω_u are 0.576 and 0.336 respectively. When an ideal relay is employed, the percentage errors in K_u and ω_u are -15.8% and -15.09% respectively. Again, improvement in the estimates of K_u and ω_u can be seen as we decrease the slope toward k_{min} (Figure 5.10). This example again shows that setting $k = 1.4 k_{min}$ will lead to quite accurate values of ultimate gain and ultimate frequency while guaranteeing the success of the relay feedback test. ∎

With these results, we can devise a procedure to find more accurate values of K_u and ω_u.

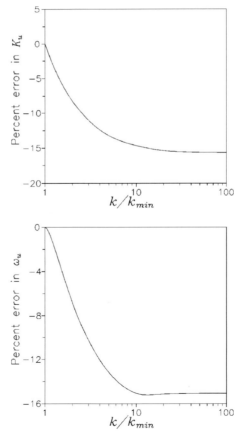

Figure 5.10. Percentage error in K_u and τ_u for Example 5.3

5.2.2 Procedure

Since k_{min} (or K_u) is needed to find the slope of the saturation relay, the proposed procedure finds a rough estimate of k_{min} first and goes on to find k and then to obtain better values of K_u and ω_u. The procedure is summarized as follows. Consider a relay feedback system.

(1) Select the height of the relay h (upper and lower limits in the manipulated input in the experiment).

(2) Perform relay feedback tests:

   (a) Use an ideal relay (set the slope of the saturation relay to a large value) to estimate $\hat{K}_u$ ($\hat{K}_u = 4h/\pi a$).

   (b) Calculate the slope of the saturation relay $k = 1.4 k_{min}$ ($k_{min} = \hat{K}_u$).

   (c) Continue the relay feedback experiment using the saturation relay with $k = 1.4 k_{min}$.

(3) Find ω_u from the relay feedback test and compute the ultimate gain from

$$K_u = \frac{2h}{\pi a}\left[\left(\sin^{-1}\frac{\bar{a}}{a}\right) + \left(\frac{\bar{a}}{a}\sqrt{1-\left(\frac{\bar{a}}{a}\right)^2}\right)\right].$$

5.3 Applications

The saturation relay feedback is applied to system identification (identifying K_u and ω_u), as well as autotuning of multivariable systems. Both a linear system and a nonlinear process are studied.

Consider the WB column studied in Example 5.1. The exact values for K_u and ω_u are 2.1 and 1.608 respectively. The proposed procedure goes as follows. The relay height h is chosen as 1. Initially, a positive change in R is made and x_D starts to increase (Figure 5.11). As soon as x_D moves upward, R is set to the lower position ($\Delta R = -1$). This ideal relay feedback test goes on for two to three cycles (e.g. time < 11 min in Figure 5.11) and we can estimate K_u from the system responses. The result is $K_u = 1.71$ (-18.6% error). With the initial result, the slope of the saturation relay is chosen as $k = 1.4 k_{min} = 1.4 \times 1.71 = 2.4$, then the relay feedback test continues with the saturation relay (e.g. time > 11 min in Figure 5.11). The results show that the ultimate gain and ultimate frequency found from the saturation relay feedback are 2.098 and 1.606 respectively. This corresponds to 0.01% error in K_u and 0.012% error in ω_u. Obviously, significant improvement can be made using the proposed procedure.

A nonlinear distillation example is used to illustrate the accuracy of the proposed autotune identification procedure. The column studied by Shen and Yu

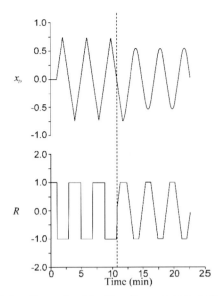

Figure 5.11. Saturation relay identification procedure for WB column

[6] is a 20-tray distillation column. The product specifications are 98% and 2% of the light component on the tops and bottoms of the column respectively. The relative volatility is 2.26, with a reflux ratio 1.76. Table 5.1 gives the steady state values. The control objective is to maintain the top and bottom product compositions by changing the reflux flow rate R and vapor boil-up rate V. This is the conventional $R-V$ control structure (Figure 5.12). This is a 2×2 system (more discussion on the multivariable system will be presented in Chapter 6). First, the $x_B - V$ loop is used to test the accuracy of the proposed method in finding K_u and ω_u. Figure 5.13 shows the input V and output x_B responses using the proposed autotune identification with a relay height of 5%. The results (Table 5.2) show that, compared with the stepping technique [1], the ideal relay feedback experiment gives significant errors in K_u and ω_u. On the other hand, the saturation relay feedback with a slope of $k = 1.4 k_{min}$ ($k = 704$) gives a very good estimate in K_u and ω_u. The errors in K_u and ω_u are -2.8% and -3.3% respectively.

Next, the saturation relay feedback is applied to both loops of this $R-V$ controlled column. The multiple-input–multiple-output (MIMO) autotuning is performed sequentially starting from the $x_D - R$ loop while keeping the $x_B - V$ loop on manual. When K_u and ω_u for the $x_D - R$ loop are found, the PI controller is tuned according to

$$K_c = \frac{K_u}{3} \tag{5.35}$$

$$\tau_I = 2 P_u \tag{5.36}$$

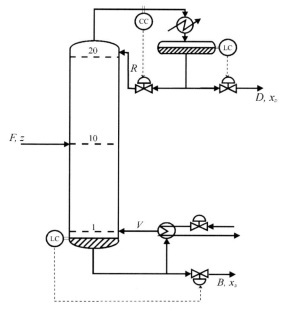

Figure 5.12. R–V controlled moderate-purity distillation column

Table 5.1. Steady state values for moderate-purity distillation column

Parameters	Values
Number of trays	20
Feed tray	10
Relative volatility	2.26
Operating pressure (atm)	1.0
Feed flow rate (kg-mole/min)	36.3
Distillation flow rate (kg-mole/min)	18.15
Bottoms flow rate (kg-mole/min)	18.15
Reflux ratio	1.76
Feed composition (mole fraction)	0.50
Distillation composition (mole fraction)	0.98
Bottoms composition (mole fraction)	0.02

With this set of tuning constants, the $x_D - R$ loop is closed and the saturation relay feedback is performed on the $x_B - V$ loop. This procedure is repeated until the tuning constants converge. Chapter 6 will discuss this MIMO autotuning procedure in detail. Figure 5.14 shows that it takes two saturation relay feedback tests to complete the autotuning procedure. In order to test the correctness of the identified system, the closed-loop transfer functions $g_{11,CL}$ obtained from different approaches are compared. Let 1 and 2 denote x_D and x_B. The closed-loop transfer function for the $x_D - R$ loop is

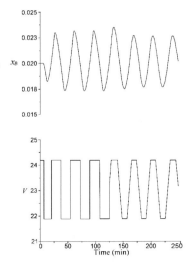

Figure 5.13. Proposed autotune identification procedure in x_B–V loop for moderate-purity distillation column

Table 5.2. Identification results (K_u and τ_u) in x_B–V loop for moderate-purity distillation column

	K_u		ω_u	
	Value	Error(%)	Value	Error(%)
Steping method	562.2	0	0.1862	0
Ideal relay ($k \to \infty$)	503.2	−10.5	0.1839	−5.2
Saturation relay ($k = 704$)	547.1	−2.8	0.1802	−3.3

$$g_{11,CL} = g_{11}\left(1 - \frac{g_{12}g_{21}}{g_{11}g_{22}} \frac{g_{22}K_2}{1+g_{22}K_2}\right) \quad (5.37)$$

Provided with the steady state gains and time constant, K_u and ω_u from the saturation relay feedback (Figure 5.14) are used to back-calculate the coefficients of $g_{11,CL}$ and $g_{22,CL}$:

$$g_{11,CL}(s) = \frac{0.00965e^{-6s}}{(9.89s+1)(23s+1)}$$

$$g_{22,CL}(s) = \frac{0.01316e^{-6s}}{(4.7s+1)(24.5s+1)}$$

The two closed-loop transfer functions are compared with the frequency responses from the stepping technique. Results (Figure 5.15) show that the saturation-relay-based MIMO autotuning gives very accurate estimates of the process transfer functions. As can be seen from Figure 5.15, it matches perfectly with the analytical results (results from stepping technique). Furthermore, autotuning based on saturation relay feedback gives satisfactory closed-loop performance for ±20% changes in feed composition (Figure 5.16).

The linear and nonlinear examples as well as identification and autotuning results clearly indicate that the saturation relay feedback gives a significant improvement in finding K_u and ω_u and, subsequently, leads to better performance in identification and MIMO autotuning.

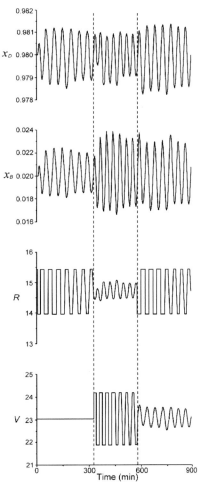

Figure 5.14. MIMO autotuning of the moderate-purity distillation example

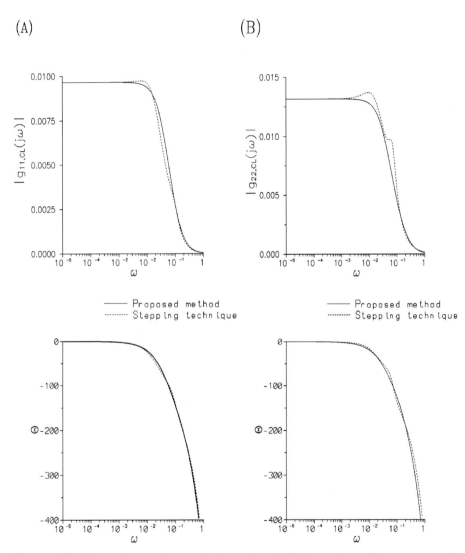

Figure 5.15. Bode diagram for moderate-purity distillation column: (A) $g_{11,CL}$ and (B) $g_{22,CL}$

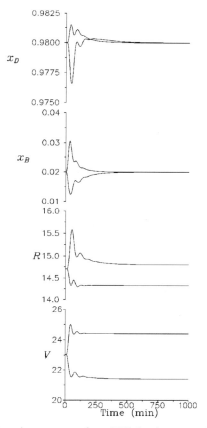

Figure 5.16. Load responses for ±20% feed composition changes

5.4 Conclusion

In this chapter we have shown that an alternative experimental design can improve the accuracy in the estimate of the ultimate gain and ultimate frequency. The analyses show that significant improvement in the estimates of K_u and ω_u can be achieved using the saturation relay feedback. It also shows that too small a slope in the ramp-type relay may fail to generate a limit cycle and, subsequently, leads to a failed experiment. A procedure is proposed to overcome the potential problem. It should be recognized that the input design plays a vital role for system identification. In addition to the saturation relay, a dual-height relay was used by Sung *et al.* [7] to approximate the sine wave.

5.5 References

1. Luyben WL. Derivation of transfer functions for highly nonlinear distillation columns. Ind. Eng. Chem. Res. 1987;26:2490.

2. Atherton DP. Nonlinear control engineering. New York: Van Nostrand Reinhold; 1982.

3. Chang RC, Shen SH, Yu CC. Derivation of transfer function from relay feedback systems. Ind. Eng. Chem. Res. 1992;31:855.

4. Li W, Eskinat E, Luyben WL. An improved autotune identification method. Ind. Eng. Chem. Res. 1991;30:1530.

5. Papastathopoulou HS, Luyben WL. Tuning controllers on distillation columns with the distillate-bottoms structure. Ind. Eng. Chem. Res. 1990;29:1859.

6. Shen SH, Yu CC. Indirect feed forward control: Multivariable systems. Chem. Eng. Sci. 1992;47:3085.

7. Sung SW, Park JH, Lee IB. Modified relay feedback method. Ind. Eng. Chem. Res. 1995;29:4133.

6
Multivariable Systems

Up to this point, discussions on autotuners are mostly limited to single-input–single-output (SISO) systems. Koivo and Pohjolainer [1] use step responses to find the state-space model for an $n \times n$ multivariable system. PI controllers are designed according to the linear model. Conceptually, this is similar to a multivariable version of the process reaction curve method. Since step responses are employed in the identification phase, the method may encounter difficulties with highly nonlinear processes. Cao and McAvoy [2] evaluate and analyze the performance of a pattern recognition controller in a multivariable system. Hsu *et al.* [3] attempt to extend the Åström–Hägglund autotuner to multivariable systems when I-only controllers are used. Furthermore, the method of Hsu *et al.* [3] requires that the steady state gain matrix should be known *a priori*. Obviously, this requirement limits the applicability of the autotuner in an operating environment.

In this chapter, we extend the Åström–Hägglund autotuner to unknown multivariable systems. Here, decentralized PI controllers are used and the square multivariable systems are assumed to be open-loop stable.

6.1 Concept

6.1.1 Single-input–Single-output Autotuning

Basically, an automatic tuning procedure can be divided into two steps: the identification phase and the controller design phase. In the identification phase, the Åström–Hägglund autotuner is based on the observation that a feedback system in which the output y lags behind the input u by $-\pi$ radians may oscillate with a period of P_u. This is a well-known observation. To generate a sustained oscillation, a relay feedback test is performed (Figure 6.1A). Initially, the input (u) is increased by h ($u = \bar{u} + h$ where $\bar{u}$ is the steady state value of u). As soon as the output is moving upward, the input is switched to the lower position ($u = \bar{u} - h$) as shown in Figure 6.1B. This procedure is repeated until the cycling has stabilized.

98 Autotuning of PID Controllers

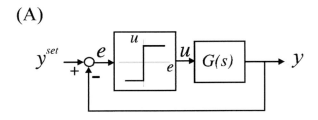

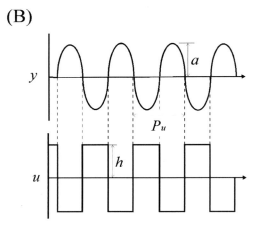

Figure 6.1. Relay feedback system: (A) block diagram and (B) response

From the relay feedback test, the familiar ultimate gain K_u and ultimate frequency ω_u are readily available. They can be approximated as

$$K_u = \frac{4h}{\pi a} \quad (6.1)$$

$$\omega_u = \frac{2\pi}{P_u} \quad (6.2)$$

where a is the amplitude of the oscillation and P_u is the period (Figure 6.1B). Following the identification step, the controller can be designed from K_u and ω_u. Typically, Ziegler–Nichols types of method can be applied to find the controller gain K_c and reset time τ_I of a PI controller. It is also possible to find the tuning constants using any other design methods, *e.g.* gain or phase margin specification [4], dominant pole design [5], M-circle criterion [6]. Notice that all these methods design a controller based on a *single* point on the Nyquist curve of $G(j\omega)$.

6.1.2 Multiple-input–Multiple-output Autotuning

In Chapter 3, the relay feedback test (ATV method) is employed for the *identification* of the transfer function matrix $G(s)$ for multivariable systems. Consider an $n \times n$ multivariable system $G(s)$ with the entry $g_{ij}(s)$. A relay feedback test is performed and each individual element in a column of the transfer function matrix can be found by fitting the corresponding point on the Nyquist curve to an assumed model. This procedure is repeated (n times) until all n^2 transfer functions in $G(s)$ are found. Once $G(s)$ is identified, the familiar *independent* (as opposed to sequential) design methods [7], *e.g.* BLT method [8], μ (structured singular value) tuning criterion [9], can be employed to find the tuning constants. By independent design, we mean all controllers are designed independently (once the performance specification is met).

Basically, these are *off-line* controller design methods, despite the fact they can be automated. However, the identification-design procedure along this line can be computationally extensive (n^2 transfer functions have to be fitted followed by an iterative searching for controller parameters). Furthermore, the columnwise identification procedure (identifying a column of transfer functions in $G(s)$ by perturbing a manipulated input) can lead to inconsistency in nonlinear multivariable processes. This is termed *independent* identification hereafter, and it will be discussed in a greater detail in the next section.

For an autotuner, an efficient identification-design procedure has to be devised. From the efficiency point of view, an important question to ask is: Do we really need to find out *all* (n^2) individual transfer functions g_{ij} for the controller design? The sequential design [7,10–14] provides an attractive alternative in MIMO autotuning. By sequential design we mean each controller (in a multivariable system) is designed in sequence. In other words, a MIMO process is treated as a sequence of SISO systems.

Based on the concept of sequential design, a simple method is proposed for MIMO autotuning. For the purpose of illustration, a 2×2 system is used (extension to an $n \times n$ system is straightforward). Consider a 2×2 system with a known pairing ($y_1 - u_1$ and $y_2 - u_2$) under decentralized control (Figure 6.2). Initially, a relay is placed between y_1 and u_1, while loop 2 is on *manual* (Figure 6.2A). Following the relay feedback test, a controller can be designed from the ultimate gain and ultimate frequency. The next step is to perform the second relay feedback test between y_2 and u_2 while loop 1 is on *automatic* (Figure 6.2B). A controller can also be designed for loop 2 following the relay feedback test. Once the controller on loop 2 is put on automatic, another relay feedback experiment is performed between y_1 and u_1 (Figure 6.2C). Generally, a new set of tuning constants is found for the controller in loop 1. This procedure is repeated until the controller settings converge. Typically, the controller parameters converge in three or four relay feedback tests for 2×2 systems. Notice that the proposed MIMO autotuning concept repeats the "identification-design" procedure on SISO transfer functions.

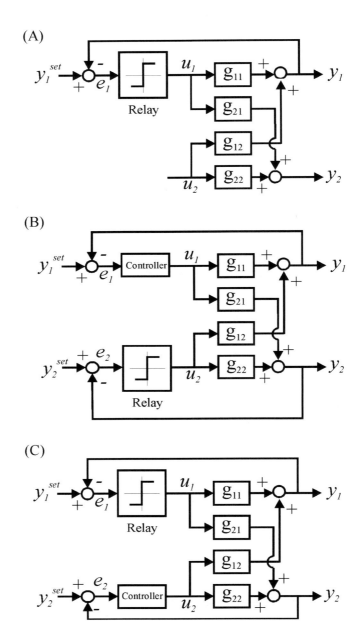

Figure 6.2. Sequential tuning procedure for a 2 × 2 system

This approach has several advantages. First, it makes the problem *simple*. The reason is that the proposed approach treats the MIMO system as a sequence of SISO systems for which the relay feedback system has been proven useful and reliable. Second, it operates in an *efficient* manner, since the autotuner identifies the transfer functions *just* needed for the controller design, as opposed to identifying and fitting all n^2 g_{ij} values in the independent design. Third, in terms of identification, it is a more accurate approach. It can be understood qualitatively that the independent identification finds linear g_{ij} values at different operating points (*e.g.* changing one input by maintaining the rest of the inputs constant) and the controller design is based on some combinations of g_{ij} values while the sequential identification finds the transfer function (this can be a combination of g_{ij} values) it needs for controller design at a single operating point. From this point of view, it is not necessary to identify all individual transfer g_{ij} values for the purpose of controllers tuning.

6.2 Theory

Since multivariable autotuning is based on the concept of a sequential identification-design procedure, the fundamental theory of sequential design is addressed. Process characteristics from the sequential design are also explored. More importantly, the idea of sequential identification is proposed and the advantages for identification in a sequential manner are also discussed.

6.2.1 Sequential Design

The concept of sequential design was proposed in the 1970s [7,10–13,15]. The idea of sequential design is a straightforward one: to treat an $n \times n$ multivariable design problem as a sequence of n SISO design problems. Therefore, the familiar design methods can be employed. Unlike the familiar multivariable design methodology in chemical process control, the classical sequential design method addresses the problems of the variable pairing and the controller design at the same time, which makes the design procedure complicated. Bhalodia and Weber [15] and Chiu and Arkun [7], on the other hand, assume that the controller structure (variable pairing) is determined *a priori* and the controller design is carried out sequentially. In this work, only the problem of controller design is addressed; the variable pairing problems are discussed elsewhere [16,17]. Notice that, in sequential design, the process transfer function matrix $G(s)$ is, generally, assumed to be known.

Consider an $n \times n$ multivariable system with decentralized PI controllers (Figure 6.3A). In the autotuning procedure, initially, the relationship between y_1 and u_1 is simply (Figure 6.2A)

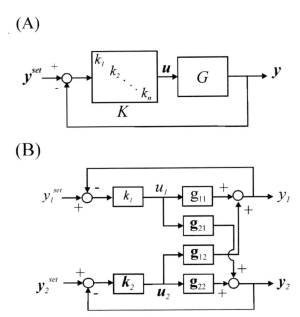

Figure 6.3. Block diagram of an n × n multivariable system with decentralized controllers

$$\left(\frac{y_1}{u_1}\right)_{OL} = g_{11} \qquad (6.3)$$

where the subscript *OL* stands for open loop. When the loop is closed sequentially, the closed-loop relationship between y_1 and u_1 becomes a bit more complicated (Figure 6.2C). The following notation will be used:

$$G(s) = \begin{pmatrix} g_{11} & \vdots & g_{12} \\ \cdots & \vdots & \cdots \\ g_{21} & \vdots & g_{22} \end{pmatrix}_{n-1}^{1}$$

$$K(s) = \begin{pmatrix} k_1 & \vdots & 0 \\ \cdots & \vdots & \cdots \\ 0 & \vdots & k_2 \end{pmatrix}$$

$$y = \begin{pmatrix} y_1 \\ \cdots \\ y_2 \end{pmatrix}_{n-1}^{1}$$

$$u = \begin{pmatrix} u_1 \\ \cdots \\ u_2 \end{pmatrix}$$

With this partitioning (Figure 6.3B), the closed-loop relationship between y_1 and u_1 becomes

$$g_{11,CL}(s) = \left(\frac{y_1}{u_1}\right)_{CL} = g_{11}\left[1 - \frac{g_{12}g_{22}^{-1}h_2 g_{21}}{g_{11}}\right] \quad (6.4)$$

where

$$h_2 = g_{22}k_2[I + g_{22}k_2]^{-1} \quad (6.5)$$

which is the complementary sensitivity function for the rest loops and the subscript CL stands for closed-loop. When y_2 is under perfect control ($h_2 = I_{(n-1)\times(n-1)}$), $g_{11,CL}(s)$ is reduced to a simpler form:

$$g_{11,CL}(s) = \left(\frac{y_1}{u_1}\right)_{CL} = g_{11}\left[1 - \left(1 - \frac{1}{\lambda_{11}(s)}\right)\right] \quad (6.6)$$

where $\lambda_{11}(s)$ is exactly the (1,1) entry of the relative gain array (RGA [18]). When y_2 is without control ($h_2 = 0$), then $g_{11,CL}$ becomes

$$g_{11,CL}(s) = g_{11}(s) \quad (6.7)$$

Therefore, the second factor in the RHS of Equation 6.4 gives the measure of closed-loop interaction throughout the frequency range of interest. In terms of sequential design, this implies, eventually, the SISO system we are dealing with is of the form of Equation 6.4.

Let us take a 2×2 system as an example. The closed-loop relationship between y_1 and u_1 becomes

$$\begin{aligned} g_{11,CL}(s) &= g_{11}\left(1 - \frac{g_{12}g_{21}}{g_{11}g_{22}} \cdot h_2\right) \\ &= g_{11}(1 - \kappa h_2) \end{aligned} \quad (6.8)$$

where κ is the Rijnsdorp interaction measure and $h_2 = g_{22}k_2/(1 + g_{22}k_2)$. Similarly, we have

$$g_{22,CL} = g_{22}(1 - \kappa h_1) \quad (6.9)$$

Here, κ takes the system interaction into account.

Therefore, the concept of sequential design is illustrated in Figure 6.4 with the redrawn block diagram. Figure 6.4 shows that both controllers k_1 and k_2 can be designed individually in a sequential manner (e.g. the controller k_2 is embedded in $g_{11,CL}$). Despite the fact that this 2×2 system can be treated as two SISO systems, the characteristics of $g_{ii,CL}$ are quite different from the familiar process transfer functions, e.g. FOPDT system.

It can be shown that the roots of the closed-loop characteristic equation

$$\det(I + GK) = 0 \quad (6.10)$$

104 Autotuning of PID Controllers

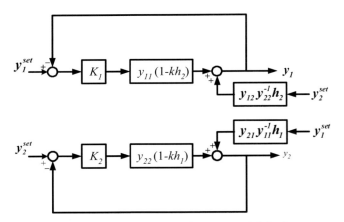

Figure 6.4. Block diagram for sequential design

are exactly the same as these for each loop in Figure 6.4.

$$1 + g_{11}(1 - \kappa h_2)k_1 = 0$$
$$\text{or} \tag{6.11}$$
$$1 + g_{22}(1 - \kappa h_1)k_2 = 0$$

The poles and zeros of the transfer functions, e.g. $g_{11,CL}$ and $g_{22,CL}$, play an important role in the controller design. Following the steps of multivariable autotuning (Figure 6.2), initially, the poles and zeros are exactly the same as those of g_{11}. However, after k_1 is designed and loop 1 is closed, the process transfer function in loop 2 becomes $g_{22}(1-\kappa h_1)$ (Figure 6.2B or 6.4). The poles of $g_{22,CL}$ are the poles of g_{22}, g_{12}, g_{21} and h_1 [13], as shown in Table 6.1. Once k_2 is designed and loop 2 is closed, we go back to loop 1, as shown in Figure 6.2C. At this point, the transfer function becomes $g_{11}(1-\kappa h_2)$ and the poles are the same as those of g_{11}, g_{12}, g_{21} and h_2. The zeros for $g_{11,CL}$ and $g_{22,CL}$ are zeros of $(1-\kappa h_2)$ and $(1-\kappa h_1)$ respectively. From the pole–zero configuration in sequential design (Table 6.1), the behavior of $g_{ii,CL}$ can be investigated.

6.2.2 Process Characteristics

Consider the autotuning steps in Figure 6.2. Initially, controller 1, k_1, is designed based on $g_{11}(s)$ using any possible SISO tuning methods. Typically, in chemical processes, $g_{ii}(s)$ is an FOPDT process transfer function, e.g. $g_{ii}(s) = K_p e^{-Ds}/(\tau_p s + 1)$. Most SISO tuning methods result in a complementary sensitivity function h_1 with a resonant peak ($L_{c,max} > 0 dB$). For example, if the Ziegler–Nichols method is used, the complementary sensitivity function h_1 shows under-damped behavior. Figure 6.5 shows the damping coefficient for the poles of h_1

Table 6.1. Poles and zeros in sequential design for a 2 × 2 system

	Poles	Zeros
Step 1 (Loop 1)	g_{11}	g_{11}
Step 2 (Loop 2)	$g_{22}, g_{12}, g_{21}, h_1$	$1 - \kappa h_1$
Step 3 (Loop 1)	$g_{11}, g_{12}, g_{21}, h_2$	$1 - \kappa h_2$

when the Ziegler–Nichols tuning is applied to FOPDT systems with a range of D/τ_p values. Notice that first-order Padé approximation is applied to find the poles of h_1. The damping coefficients fall between 0.4 to 0.5 for a range of D/τ_p (0.001–1), rather underdamped behavior. Figure 6.6 shows $L_{c,max}$ for h_1 when the Ziegler–Nichols tuning rule is applied to the same system. Again, underdamped behavior is observed.

The next step is to design k_2 when loop 1 is closed. As shown in Table 6.1, the poles of $g_{22,CL}$ are the poles of h_1, g_{22}, g_{12} and g_{21}. Therefore, $g_{22,CL}$ has a pair of *underdamped* poles (from the poles of h_1). This is a rather unusual situation, since most SISO tuning methods deal with an overdamped transfer function (*e.g.* FOPDT system). Actually, chemical processes rarely show underdamped open-loop responses (*e.g.* considering separators and reactors). Here, the underdamped characteristics have resulted from the sequential design of multivariable systems.

Since the multivariable system is treated as a series of SISO systems (Figure 6.4), a form of process transfer function for $g_{ii,CL}$ is helpful for the purpose of analysis. In addition to the underdamped poles, the pole of g_{ii} is also the pole of $g_{ii,CL}$. Therefore, an approximate transfer function is used:

$$g_{ii,CL}(s) = \frac{K_p}{\left(\tau_p^2 s^2 + 2\tau_p \zeta s + 1\right)} \left(\frac{\tau_{p2} s + 1}{\tau_{p1} s + 1}\right) e^{-Ds} \qquad (6.12)$$

This is a rather unusual structure for a typical process transfer function. However, it gives a description of the mix of underdamped and overdamped behavior which resulted from sequential design. A 2×2 distillation column example is used to illustrate the appropriateness of Equation 6.12.

Example 6.1 WB column
Consider the transfer function matrix

$$\begin{pmatrix} y_1 \\ y_2 \end{pmatrix} = G(s) \begin{pmatrix} u_1 \\ u_2 \end{pmatrix} = \begin{pmatrix} \dfrac{12.8 e^{-Ds}}{16.7s + 1} & \dfrac{-18.9 e^{-3s}}{21s + 1} \\ \dfrac{6.6 e^{-7s}}{10.9s + 1} & \dfrac{-19.4 e^{-3s}}{14.4s + 1} \end{pmatrix} \begin{pmatrix} u_1 \\ u_2 \end{pmatrix} \qquad (6.13)$$

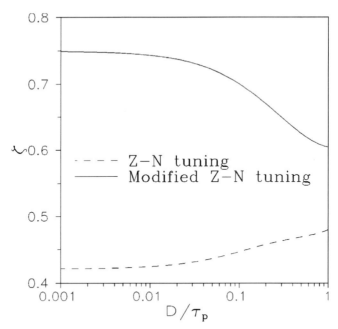

Figure 6.5. Damping coefficient for an FOPDT system with original and modified Ziegler–Nichols methods

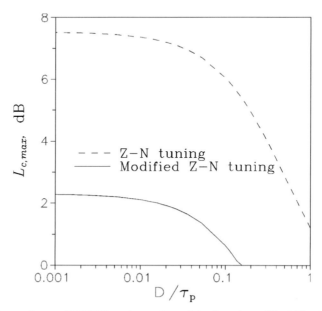

Figure 6.6. $L_{c,max}$ for an FOPDT system with original and modified Ziegler–Nichols methods

with the tuning constants for PI controllers: $K_{c1} = 0.54$, $K_{c2} = -0.072$ and $\tau_{I1} = 7.92$, $\tau_{I2} = 26.7$. The underdamped step response of loop 1 ($g_{11,CL} = g_{11}(1-\kappa h_2)$) is shown in Figure 6.7. The step responses data are fitted to Equation 6.12. The results of least square regression give

$$\hat{g}_{11,CL}(s) = \frac{6.4}{42.25s^2 + 11.7s + 1}\left(\frac{44s+1}{60s+1}\right)e^{-s}$$

Figure 6.7 compares the step responses of the original process and the approximated model $\hat{g}_{11,CL}$. Good approximation can be obtained using Equation 6.12. ∎

Another characteristic of $g_{ii,CL}(s)$ comes from the zeros. Table 6.1 shows that the zeros of $g_{22,CL}$ are those of $(1-\kappa h_1)$. Consider a case of $\kappa(0) > 1$ (i.e. the system is paired with negative RGA in the diagonal, $\lambda_{ii} < 0$). It then becomes obvious that RHP zero *can* occur. For example, if we have

$$h_1(s) = \frac{1}{s+1}$$

and

$$\kappa(s) = 5$$

The zero of $g_{22,CL}$ is 4 (an RHP zero). However, if

$$\kappa(s) = 5(s+1)^2$$

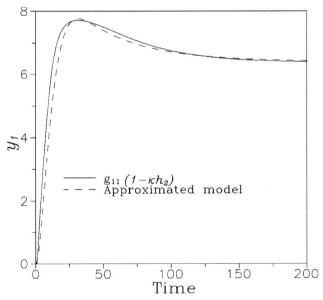

Figure 6.7. Step responses of the original process $g_{11}(1 - kh_2)$ and the approximated model (Equation 6.12)

then the zero becomes $-4/5$ (an LHP zero). This confirms the finding that pairing with a negative RGA does not necessarily result in inverse responses (steady state information is not sufficient to decide [16]). Nonetheless, systems with $\kappa(0) > 1$ give a different sign in the controller gain from the open-loop point of view.

6.2.3 Sequential Identification

It is well understood that system identification plays an important role for the success of an autotuner. Traditionally, identification of MIMO systems is carried out by manipulating the inputs u_i *independently*. That is, the first column of the transfer function matrices $g_{i1, i=1,...,n}$ are obtained for a change in u_1 while the rest of the inputs $u_{j, j \neq 1}$ are kept constant. Figure 6.8A illustrates the signal flow in the independent identification.

However, difficulties arise for the identification of nonlinear multivariable processes [19,20]. Despite the fact that the errors for each individual transfer function $g_{ij}(s)$ are at an acceptable level, the identified transfer function matrices simply fail to describe fundamental process characteristics. For example, Luyben [21] shows that in order to find the correct RGA (in the sign), the changes made in the manipulated input u_i are so small (0.05% changes) that the computer calculations

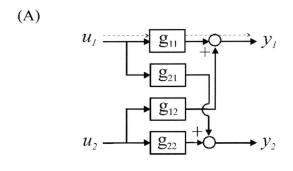

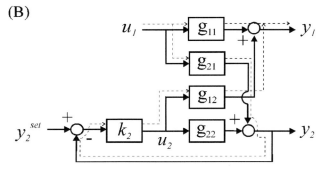

Figure 6.8. Signal flow in (A) independent identification and (B) sequential identification

have to be carried out using double precision (not to mention how to implement it in an operating environment).

In their pioneering work, Häggblom and Waller [22] point out the problem: consistency relations are not met for individual transfer functions. In a series of papers, Waller and coworkers utilize "external material balances" (consistency relations) to find the transformations between control structures, reconcile process models and design controllers for disturbance rejection, *etc.* Häggblom and Waller [23] give a good summary. Notice that, in their work, independent identification is performed (or assumed) and then the consistency relations are enforced. The goal for all the reconciliation is obvious: the elements in a process transfer function matrix should follow some sort of consistency relations (*e.g.* satisfying material balances). A new approach is proposed to achieve this goal by *modifying* the identification procedure.

In designing controllers for a multivariable system, the actual transfer function we need is, generally, a combination of g_{ij} values. For example, in the sequential design for a 2×2 system the actual transfer function we need is

$$g_{ii,CL}(s) = g_{ii}\left(1 - \frac{g_{12}g_{21}}{g_{11}g_{22}} \cdot h_j\right), \quad i = 1, 2 \text{ and } j \ne i \tag{6.14}$$

If these g_{ij} values come directly from independent identification (without checking consistency relations), then the design can be erroneous. A simple way to meet consistency relations is by performing the identification in a sequential manner: sequential identification (Figure 6.8B). Figure 6.2 illustrates the procedure of sequential identification when the relay feedback test is employed.

The advantage of sequential identification is shown in the following example and comparisons are made between independent and sequential identifications.

Example 6.2 Blending system
Consider a simple 2×2 blending system (Figure 6.9). The control objective is to maintain flow rate in the outlet stream F using the first stream F_1 and the composition is controlled by changing the second stream F_2. Material balances describing the blending system are

$$F = F_1 + F_2 \tag{6.15}$$

$$xF = x_1 F_1 + x_2 F_2 \tag{6.16}$$

Linearizing Equation 6.16, the process transfer function $G(s)$ describing this nonlinear system becomes

$$\begin{pmatrix}\Delta F \\ \Delta x\end{pmatrix} = \begin{pmatrix} 1 & 1 \\ \dfrac{(\overline{x}_1 - \overline{x})}{\overline{F}} & \dfrac{(\overline{x}_2 - \overline{x})}{\overline{F}} \end{pmatrix}\begin{pmatrix}\Delta F_1 \\ \Delta F_2\end{pmatrix} \tag{6.17}$$

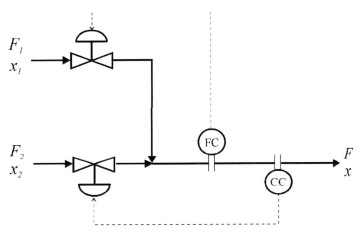

Figure 6.9. Blending system

The nominal steady state conditions are $\bar{x} = 0.78, \bar{F} = 20, \bar{x}_1 = 0.9, \bar{F}_1 = 16, \bar{x}_2 = 0.3, \bar{F}_2 = 4$. For independent identification, F_1 and F_2 are each perturbed by a factor of 50%. Notice that the results from the step changes in F_1 and F_2 fail to satisfy the component material balance (the consistency relation):

$$\bar{x}\Delta F + \Delta x \bar{F} = \bar{x}_1 \Delta F_1 + \bar{x}_2 \Delta F_2 \tag{6.18}$$

Table 6.2 gives the values of g_{ij} from independent identification. Results show that g_{21} and g_{22} deviate from the true value by −28.3% and −9.2% respectively. Obviously, the errors depend on the magnitude of the step changes. Furthermore, the resultant $g_{ii,CL}$ values are also quite different from the true values (Table 6.2). On the other hand, sequential identification (Figure 6.8B) can find $g_{ii,CL}$ directly. Step changes of 50% are made on u_1 and u_2, sequentially, while the other loop is closed (Figure 6.8). Notice that sequential design finds $g_{ii,CL}$ values directly (bypassing g_{ij}), as shown in Table 6.2. The $g_{ii,CL}$ values found are exactly the same as the true values (Table 6.2). ∎

This example clearly shows the advantage of the proposed identification approach for nonlinear multivariable systems. The sequential identification finds the *essential* element, $g_{ii,CL}$, for controller design. In doing this, the consistency relations are achieved *internally*.

Table 6.2. Estimated process transfer functions for different identification approaches

	g_{11}	g_{21}	g_{12}	g_{22}	$g_{11,CL}$	$g_{22,CL}$
True values	1	0.0060	1	−0.0250	1.250	−0.030
Independent identification	1	0.0043	1	−0.0218	1.197	−0.026
Sequential identification	−	−	−	−	1.250	−0.030

6.3 Controller Tuning

6.3.1 Potential Problem in Ziegler–Nichols Tuning

The Ziegler–Nichols method [24] is a very popular method of tuning the PID type of controller for the reasons of its simplicity and its experimental nature (an experimental procedure comes with the tuning rule). However, most studies of the Ziegler–Nichols method deal with *overdamped* systems [5,25,26]. The stability problem may arise when one tries to tune an underdamped system using the Ziegler–Nichols method. Tan and Weber [27] explored stability problems associated with Ziegler–Nichols tuning for third-order systems. As pointed out earlier, the sequential design may produce an underdamped system since the poles of h_j are also the poles of $g_{ii,CL}$ ($i \neq j$), and a typical process transfer function is given in Equation 6.12. Let us use the transfer function of Equation 6.12 to illustrate the stability problem in the Ziegler–Nichols method. Consider the following underdamped system with a damping coefficient of 0.6:

$$g(s) = \frac{1}{25s^2 + 6s + 1}\left(\frac{10s+1}{5s+1}\right)e^{-s}$$

Based on the Ziegler–Nichols method, the controller settings are $K_c = 19.39$ and $\tau_I = 2.84$. The Nyquist plot of GK shows that the closed-loop system is unstable (Figure 6.10). From Figure 6.10, it is clear that for an underdamped system ($\zeta = 0.6$ in this example) the Ziegler–Nichols method may give an unstable closed-loop system. Notice that the results of Figure 6.5 show that, for the most common type of transfer function (FOPDT), the Ziegler–Nichols method produces underdamped poles (poles of h) with the damping coefficient ranging from 0.4 to 0.5. Furthermore, in the sequential design, the underdamped transfer function, *i.e.* $g_{ii,CL}$, has to be tuned *again* with the simple Ziegler–Nichols method. This will lead to an even more underdamped closed-loop system. Therefore, modifications have to be made to avoid underdamped poles.

6.3.2 Modified Ziegler–Nichols Method

It should be emphasized that any familiar SISO tuning methods, *e.g.* gain margin, phase margin, $L_{c,max}$ criterion, can be applied to the PI controller design. However, based on the relay feedback type of identification, the Ziegler–Nichols type of method is a natural choice (since K_u and ω_u are available). It is clear that any modification should make the tuning constants more conservative. The detuning procedure follows the spirit of BLT [8]. That is, a single detuning factor is employed to find appropriate constants.

$$K_c = \frac{K_{c,ZN}}{f}$$

$$\tau_I = \tau_{I,ZN} \cdot f$$

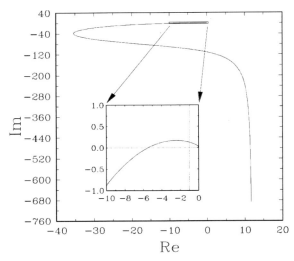

Figure 6.10. Nyquist plot of GK with Ziegler–Nichols tuning

After a number of tests on linear distillation column models [8], 2×2 systems with FOPDT transfer functions [28] and nonlinear distillation examples, a detuning factor $f \approx 2.5$ is proposed. Justifications for the proposed tuning rule will be given shortly. The modified Ziegler–Nichols method for PI controller becomes

$$K_c = \frac{K_u}{3} \qquad (6.19)$$

$$\tau_I = 2P_u \qquad (6.20)$$

The original Ziegler–Nichols PI tuning rule moves the crossover point $(-1/K_u, 0)$ in the G-plane to the point $(-1/2.2, 0.087)$ (for the same frequency ω_u) in the GK-plane. In the modified method a more conservative measure is taken, and the point corresponding to ω_u is moved to $(-1/3, 0.0265)$ in the GK-plane. Since the damping coefficient of $g_{22,CL}$ (Equation 6.14) comes from h_1, we are interested in the damping coefficient or $L_{c,\,max}$ of the proposed method when applied to typical g_{ii} values.

Figure 6.5 shows that damping coefficient of the modified Ziegler–Nichols method for the transfer functions of $e^{-Ds}/(\tau_p s + 1)$ type. The results (Figure 6.5) show that the proposed method is less underdamped (with ζ greater than 0.6 for a range of D/τ_p values). The $L_{c,\,max}$ plots in Figure 6.6 also show the same trend. This implies that, for the ranges of parameters ($0.001 \leq D/\tau \leq 1$) studied, the damping coefficient in Equation 6.14 is greater than 0.6.

Stability is an important concern for any tuning method. However, a trade-off between performance and the stability region has to be made. As pointed out by Tan and Weber [27], unstable regions can always be found for different values of damping coefficient for third-order processes. Regions of instability are investigated for the transfer function of the form

$$G(s) = \frac{K_p e^{-Ds}}{\tau_p^2 s^2 + 2\tau_p \zeta s + 1} \left(\frac{\tau_{p2} s + 1}{\tau_{p1} s + 1} \right) \qquad (6.21)$$

The following parametric spaces are studied: $K_p = 1$, $\tau_p = 5$, $\tau_{p1} = 0\text{--}10$, $\tau_{p2} = 0\text{--}10$, $\zeta = 0.1\text{--}1$. Figure 6.11A and B shows that instability regions exist for both the original and modified Ziegler–Nichols methods. However, the modified Ziegler–Nichols method reduces the instability region significantly. Figure 6.11 reveals that the instability region often occurs in the region when $\tau_{p2} > \tau_{p1}$. The reason is that the larger lead time constant in Equation 6.21 results in a resonant peak in $g(j\omega)$ which can be viewed as an enhancement of the underdamped behavior. Obviously, the instability can be eliminated from Figure 6.11 by detuning the controller further (using a much larger f). However, performance will deteriorate.

Probably the most important evaluation is to test these methods in a sequential design environment. A number of 2×2 systems are studied. Consider 2×2 systems $G(s)$ with FOPDT transfer function $g_{ij}(s)$. It is assumed that: $\lambda(s) = \lambda(0)$ (or $\kappa(s) = \kappa(0)$) and $g_{11}(s) = g_{22}(s)$. Sequential design is applied to the system with different values in of κ and D/τ_p. The maximum closed-loop log modulus $L_{c,max}$ of the complementary sensitivity function ($g_{ii,CL} = g_{ii,CL} k_i /(1 + g_{ii,CL} k_i)$) is found. Notice that typically $L_{c,max} = +2\text{dB}$ is an often used heuristic in SISO tuning [26]. Figure 6.12A shows that $L_{c,max}$ is ranging from 1.2 to 28 dB for the Ziegler–Nichols method with $\lambda > 1$ and $0.001 < D/\tau_p < 1$. Furthermore, the Ziegler–Nichols method produces an unstable system for $\lambda < 0.3$ (Figure 6.12A).

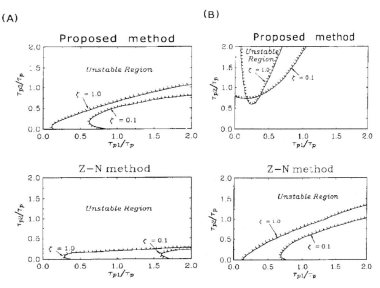

Figure 6.11. Contour plots of stability regions for different tuning methods with (A) $D = 0.1$ and (B) $D = 1$

On the other hand, the proposed method gives fairly constant $L_{c,max}$ (ranging form 0 to 5.7 dB) for $\lambda > 1$ (Figure 6.12B) and $0.001 < D/\tau_p < 1$. As λ falls below unity, the $L_{c,max}$ increases. However, the values of $L_{c,max}$ are still acceptable for $\lambda > 0.3$. For $\lambda < 0.25$, Figure 6.12B shows that an unstable region appears. Certainly, a more conservative tuning method (using a larger f) can be used to eliminate the unstable region. However, this can produce sluggish responses for systems with $\lambda > 1$ whenever the constant f tuning rule is applied.

(A)

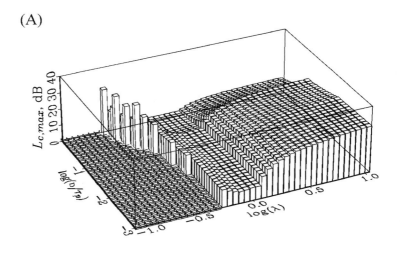

(B)

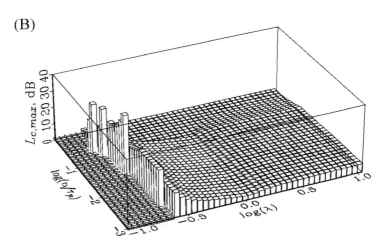

Figure 6.12. $L_{c,max}$ for 2 × 2 systems using different tuning methods (A) Ziegler–Nichols tuning and (B) modified Ziegler–Nichols tuning

The stability and $L_{c,max}$ analyses for SISO and more realistic MIMO systems show that the modified Ziegler–Nichols rule works for a typical process transfer function provided with reasonable variable pairing. More importantly, the tuning constants follow directly from the relay feedback test (with very little computation).

6.3.3 Performance Evaluation: Linear Model

The WB column example (Example 6.1) is used to test the performance of the proposed tuning method. The identification-design procedure is carried out in the 2×2 system. Except for the original pairing $y_1 - u_1$ and $y_2 - u_2$, no prior knowledge about the system is assumed. The autotuning procedure is:

(1a) perform the relay feedback test on $y_1 - u_1$ while loop 2 is on manual (Figure 6.2A)

(1b) design the PI controller k_1 based on k_{u1} and ω_{u1} according to Equations 6.19 and 6.20

(2a) perform the relay feedback test on $y_2 - u_2$ while loop 1 is on automatic (Figure 6.2B)

(2b) design the PI controller k_2

(3a) perform the relay feedback test on $y_1 - u_1$ while loop 2 is on automatic (Figure 6.2C)

(3b) design the PI controller k_1.

This completes the identification-design procedure. Figure 6.13 shows the autotuning procedure, which is completed in the first 100 min. Notice that, in theory, we need another step in the autotuning: redesign k_2 while the new set of PI tuning constants is available. However, simulation results show that a new k_1 (or h_1) has little effect on k_2. Therefore, the autotuning is terminated in three steps. A load dis-turbance is introduced at $t = 400$ min; the results show that the automatically designed controllers possess a good disturbance rejection capability (Figure 6.13). For a closer look at the performance, the modified Ziegler–Nichols method is compared with the BLT method and the empirical method [8]. Simulation results show that the load performance of the proposed tuning method is as good as other well-known tuning methods (Figure 6.14). It should be emphasized, again, that this good performance is achieved with essentially no prior knowledge about the process (with the BLT or the empirical method it is necessary to know the process transfer function for the tuning) and very little engineering effort (finding K_u and ω_u and, subsequently, K_c and τ_I from Equations 6.19–6.20).

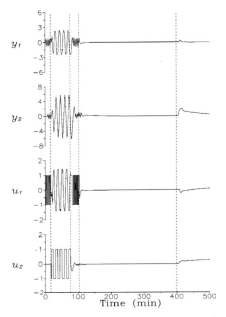

Figure 6.13. Automatic tuning and load responses for WB column

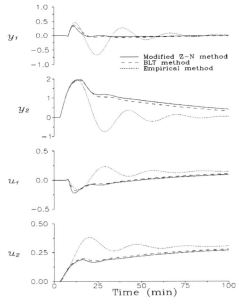

Figure 6.14. Load responses for WB column with different tuning methods

6.4 Properties

Despite the apparent success of the proposed automatic tuning method, potential problems of the proposed autotuner are raised which are helpful for the applications of the autotuner in an operating environment.

6.4.1 Convergence

In theory, the property of convergence of any iterative procedure is very important. Since the sequential identification-design procedure in MIMO autotuning is an iterative process, the convergence of the proposed autotuning procedure is discussed. In studying 2×2 systems, Bhalodia and Weber [15] point out that, starting from different loops, the sequential design converges to the same set of tuning constants. Recall the autotuning steps (e.g. steps 2 and 3 in Section 6.3.3) that, in step 2, the identification phase finds K_{u2} and ω_{u2} while loop 1 is on automatic ($h_2(j\omega)$ is known) and the controller design phase calculates k_2, consequently h_2, from K_{u2} and ω_{u2} (Equations 6.19 and 6.20). When going back to loop 1 in step 3, the purpose is to find K_{u1} and ω_{u1} with h_2 (or k_2) available (found from the previous step). Therefore, mathematically, the problem can be formulated as find ω_{u1} and ω_{u2} such that the following two nonlinear equations converge:

$$f_1(\omega_{u1},\omega_{u2}) = \tan^{-1}\left[\frac{\text{Im}\{g_{11,CL}(j\omega_{u1},j\omega_{u2})\}}{\text{Re}\{g_{11,CL}(j\omega_{u1},j\omega_{u2})\}}\right] = -\pi \tag{6.22}$$

$$f_2(\omega_{u1},\omega_{u2}) = \tan^{-1}\left[\frac{\text{Im}\{g_{22,CL}(j\omega_{u1},j\omega_{u2})\}}{\text{Re}\{g_{22,CL}(j\omega_{u1},j\omega_{u2})\}}\right] = -\pi \tag{6.23}$$

where

$$g_{11,CL}(j\omega_{u1},j\omega_{u2}) = g_{11}(j\omega_{u1})\left[1-\kappa(j\omega_{u1})\frac{g_{22}(j\omega_{u1})k_2}{1+g_{22}(j\omega_{u1})k_2}\right] \tag{6.24}$$

with

$$k_2 = \frac{1}{3|g_{22,CL}(j\omega_{u2})|}\left[1+\frac{1}{\frac{4\pi}{\omega_{u2}}j\omega_{u1}}\right] \tag{6.25}$$

And a similar expression can found for $g_{22,CL}(j\omega_{u1},j\omega_{u2})$.

Unlike the conventional way of solving this set of nonlinear equations simultaneously, Equations 6.22 and 6.23 are solved sequentially. That is, in the kth iteration Equation 6.22 is solved for ω_{u1} ($\omega_{u1}^{(k)}$) with ω_{u2} taking constant values from the previous iteration ($\omega_{u2} = \omega_{u2}^{(k-1)}$). In the *linear equations* counterpart, this is exactly the Gauss–Seidel method (see Rice [29] p.142) for solving linear algebraic equations in a sequential manner. Consider a set of linear algebraic equations

$$Ax = B \tag{6.26}$$

where A is the coefficient matrix with the entry a_{ij}, x is the solution vector and B is a vector with constant values. When solving the equation sequentially, the necessary and sufficient condition for the equation solving to converge is (see Rice [29] p.144)

$$\rho\left[I - A_{diag}^{-1} A\right] < 1 \qquad (6.27)$$

where $\rho(\cdot)$ is the spectral radius (the largest absolute value of the eigenvalue) of $(\cdot)$ and A_{diag} is the matrix with a_{ii} in the diagonal and zero elsewhere. For a 2×2 system, Equation 6.27 is equivalent to

$$\frac{a_{12} a_{21}}{a_{11} a_{22}} < 1 \qquad (6.28)$$

As for the case of sequential identification-design, the problem can be formulated as

$$\begin{pmatrix} \left(\frac{\partial f_1}{\partial \omega_{u1}}\right)_{\bar{\omega}u2} & \left(\frac{\partial f_1}{\partial \omega_{u2}}\right)_{\bar{\omega}u1} \\ \left(\frac{\partial f_2}{\partial \omega_{u1}}\right)_{\bar{\omega}u2} & \left(\frac{\partial f_2}{\partial \omega_{u2}}\right)_{\bar{\omega}u1} \end{pmatrix} \begin{pmatrix} \omega_{u1} \\ \omega_{u2} \end{pmatrix} = \begin{pmatrix} -\pi \\ -\pi \end{pmatrix} \qquad (6.29)$$

where the overbar stands for the solution of the nonlinear equations. Therefore, to check for convergence is equivalent to finding whether

$$\frac{\left(\frac{\partial f_1}{\partial \omega_{u2}}\right)_{\bar{\omega}u1} \left(\frac{\partial f_2}{\partial \omega_{u1}}\right)_{\bar{\omega}u2}}{\left(\frac{\partial f_1}{\partial \omega_{u1}}\right)_{\bar{\omega}u2} \left(\frac{\partial f_2}{\partial \omega_{u2}}\right)_{\bar{\omega}u1}} < 1 \qquad (6.30)$$

An experiment is carried out to test the convergence of the multivariable autotuning procedure. 10,000 cases of 2×2 systems with FOPDT types of transfer function are generated *randomly* and Equations 6.22 and 6.23 are solved simultaneously. Then, the condition for convergence (Equation 6.30) is checked by linearizing Equations 6.22 and 6.23 numerically. Results show that all cases meet the convergence criterion. Therefore, it can be conjectured that there is no convergence problem associated with the automatic tuning procedure around the neighborhood of the true solution. Actually, this can be understood physically. From Equation 6.24, it can be seen that the only way ω_{u2} can affect $g_{11,CL}(j\omega_{u1}, j\omega_{u2})$ is through k_2 (or h_2 in a more general way). However, generally, the magnitude of h_j is kept constant for a range of frequencies and the bandwidth of h_j often is larger than that of g_{ii}. Therefore, the change in ω_{u2} has little impact of the solution of $g_{11,CL}$. Since the systems (systems to be autotuned) are virtually unknown, the property of convergence can only be conjectured or interpreted qualitatively. As for our own experience, we have not found any convergence problem in any of the cases studied.

6.4.2 Tuning Sequence

Tuning sequence (which loop to be tuned first) is a problem which has arisen from the sequential design. The work of Bhalodia and Weber [15] implies that a different tuning sequence may result in a different speed of convergence (to the final controller parameters). In our experiments, this means some understanding of the tuning sequence can lead to an efficient identification-design procedure. In other words, if a tuning sequence can result in faster convergence, then the autotuning procedure can be terminated sooner. In our problem setting, the process is assumed to be virtually unknown except for some very qualitative description. For example, if information about the relative speed of the loop responses is available, then one may utilize this to devise an appropriate tuning sequence. The relative loop speed of Marino-Galarraga et al. [28] is useful in this regard. The normalized loop speed for the loop i is defined as [28]

$$s_i = \frac{\omega_{ui}}{\sum_{i=1}^{n} \omega_{ui}} \tag{6.31}$$

where ω_{ui} is the ultimate frequency of g_{ii}. From the definition, it is clear that s_i falls between 0 and 1. For a 2×2 system, if the speed of loop 1 (g_{11} to be exact) is faster than that of loop 2, then $s_1 > 0.5$. Notice that s_i is the relative loop speed when only diagonal elements are considered. Such information does not give a complete description of a multivariable system in a rigorous way. However, this is probably the most easily available dynamic information from the plant personnel. Once s_i is available, which loop should be tuned first?

Following the approach of Marino-Galarraga et al. [28], a simple experiment is performed. Consider 14 2×2 systems with λ_{11} values of 0.5 and 2.0 (Table 6.3). The proposed automatic tuning procedure is applied to these 14 systems starting from loop 1 and loop 2. Table 6.3 gives the model structure and the parameter values for these systems. The iterations in the sequential design are terminated when the controller parameters are within 1% of the true values. The results (Table 6.3) indicate the general trend: faster convergence can be achieved when the *fast* loop is tuned *first*.

Actually, this result can be understood physically. For a system with a very different loop speed, dynamically, the system interaction has little impact on the fast loop. For example, the effect from system interaction (through the slow loop) does not show on the fast variable until the transient (results from the transfer function of the fast loop) almost dies out. On the other hand, the effect of the fast loop always acts on the slow loop. Mathematically, the bandwidth of the complementary sensitivity function for the slow loop (e.g. bandwidth of h_2) is much smaller than that of the fast loop (e.g. h_1) and typically h_2 is a low-pass filter. Therefore, it is very likely that

$$g_{11,CL}(j\omega_{u1}) = g_{11}(j\omega_{u1})(1 - \kappa(j\omega_{u1})h_2(j\omega_{u1})) \approx g_{11}(j\omega_{u1}) \tag{6.32}$$

120 Autotuning of PID Controllers

Table 6.3. Convergence for systems with different relative speeds.

System	RGA	D_{11}	D_{12}	D_{21}	D_{22}	s_1	Iteration No. Loop 1 first/ Loop 2 first
1	0.5	0.40	2.70	3.00	5.70	0.905	2/3
2	0.5	1.00	2.70	3.00	5.70	0.811	3/4
3	0.5	2.00	2.70	3.00	5.70	0.709	3/3
4	0.5	3.00	2.70	3.00	5.70	0.633	4/5
5	0.5	4.50	2.70	3.00	5.70	0.550	3/8
6	0.5	5.00	2.70	3.00	5.70	0.527	3/8
7	0.5	6.00	2.70	3.00	5.70	0.489	4/3
8	2.0	0.40	2.70	3.00	5.70	0.905	3/4
9	2.0	1.00	2.70	3.00	5.70	0.811	3/3
10	2.0	2.00	2.70	3.00	5.70	0.709	3/3
11	2.0	3.00	2.70	3.00	5.70	0.633	2/3
12	2.0	4.50	2.70	3.00	5.70	0.550	4/5
13	2.0	5.00	2.70	3.00	5.70	0.527	4/5
14	2.0	6.00	2.70	3.00	5.70	0.489	5/3

where:

$$G(s) = \begin{pmatrix} \dfrac{K_{p11}}{\tau_{p11}s+1}e^{-D_{11}s} & \dfrac{K_{p12}}{\tau_{p12}s+1}e^{-D_{12}s} \\ \dfrac{K_{p21}}{\tau_{p21}s+1}e^{-D_{21}s} & \dfrac{K_{p22}}{\tau_{p22}s+1}e^{-D_{22}s} \end{pmatrix}$$

$\tau_{p11} = \tau_{p12} = \tau_{p21} = \tau_{p22} = 1$
for $\lambda=0.5$, $K_{p11} = K_{p12} = K_{p22} = 1$ and $K_{p21} = -1$ and
for $\lambda=2.0$, $K_{p11} = K_{p12} = K_{p22} = 1$ and $K_{p21} = 0.5$

Equation 6.32 indicates that if loop 1 is tuned first we are closer to the solution (of Equations 6.22 and 6.23). The analyses show that when qualitative information about relative loop speed is available we are able to utilize this in deciding the tuning sequence, which will lead to a more efficient autotuning procedure.

6.4.3 Problem of Variable Pairing

In theory, variable pairing should not pose any problem at this stage. That is, all outputs and manipulated inputs are paired correctly and the process is controlled via DCS or single-station controllers. However, the proposed autotuning procedure can be used to eliminate undesirable variable pairing. This is helpful at the commissioning stage of a new control system, since we are able to spot potential problems in the plant. The WB column example is used to illustrate the peculiar behavior and poor performance when an undesirable pairing is configured. Notice that the RGA for the correct pairing is $\lambda_{11}(0) = 2.01$ and the well-known fact that a system paired with negative λ_{11} is undesirable [16,18].

Example 6.3 WB column with undesirable pairing
If the WB column is paired incorrectly ($y_1 - u_2$ and $y_2 - u_1$), then we have

$$\begin{pmatrix} y_1 \\ y_2 \end{pmatrix} = G' \begin{pmatrix} u_2 \\ u_1 \end{pmatrix} = \begin{pmatrix} \dfrac{-18.9e^{-3s}}{21s+1} & \dfrac{12.8e^{-s}}{16.7s+1} \\ \dfrac{-19.4e^{-3s}}{14.4s+1} & \dfrac{6.6e^{-7s}}{10.9s+1} \end{pmatrix} \begin{pmatrix} u_2 \\ u_1 \end{pmatrix} \qquad (6.33)$$

This pairing gives a negative value in λ_{11} ($\lambda_{11} = -1.01$). Consider two cases. The first case is that we do not know the signs of the diagonal elements (in reality, it is nearly impossible not to know the signs of the diagonal elements once the control hardware is installed) or, more likely, we forget to apply the knowledge of the "sign" in the relay feedback tests. Figure 6.15 illustrates the process of the autotuning. A relay feedback test ($t = 0$–50 min in Figure 6.15) is performed on loop 1 (the $y_1 - u_2$ loop) first and a PI controller is designed using the modified Ziegler–Nichols method; then, another test is carried out on the $y_2 - u_1$ loop ($t = 50$–200 in Figure 6.15) and the second PI controller is designed. The procedure iterates back to loop 1 with a controller design and the autotuning process is terminated at $t = 300$ min, as shown in Figure 6.15. After the transient dies out, a unit step load change is applied (at $t = 400$ min). Poor load responses are observed (Figure 6.15) when compared with the load responses in Figure 6.13 (the load responses with the correct pairing).

With a closer look at the sustained oscillations when loop 2 ($y_2 - u_1$) is tuned (Figure 6.15) one can find that the closed-loop gain $g_{22,CL}$ is *negative*, which is different from the *positive* open-loop gain (the (2,2) element in Equation 6.33). Notice that if the sign of the process gain is positive, then the output and the input move toward the opposite direction in a half period (*e.g.* Figure 6.1B). Therefore, the controller gain is negative in loop 2.

$$K(s) = \begin{pmatrix} -0.123 \left(\dfrac{30.7s+1}{30.7s} \right) & 0 \\ 0 & -0.17 \left(\dfrac{78.1s+1}{78.1s} \right) \end{pmatrix} \qquad (6.34)$$

That implies that loop 2 cannot be stabilized by itself [16,17]. Despite the fact that the overall closed-loop system is stable, as shown in Figure 6.15, this control structure lacks integrity. Furthermore, $g_{11,CL}$ is (open-loop) unstable, as shown in Appendix A. Therefore, poor responses can be expected for such a poorly conditioned closed-loop system. However, no indication of any error is observed in the autotuning process (Figure 6.15), apart from the poor responses.

Let us consider the second case, where we know or are aware of the "sign" of the diagonal elements. Loop 1 ($y_1 - u_2$ loop) is tested and the controller is designed in the same way as that of the previous case. After loop 1 is closed, a second relay feedback test is performed between y_2 and u_1. An increase in u_1 is

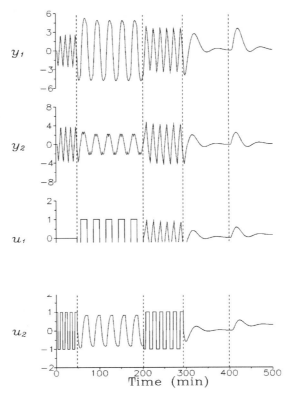

Figure 6.15. Automatic tuning procedure for the WB column with wrong pairing

made initially. Since we know the sign between y_2 and u_1 is positive, we are waiting for y_2 to cross the SP. However, y_2 simply levels off toward the negative direction and the relay never switches [30]. This means the relay feedback test fails (because we insist that the sign of the transfer function is positive). Once such a situation occurs, an undesirable pairing is confirmed. ∎

From the ongoing analyses, it is clear that if the signs of the diagonal elements are known or, most likely, one is aware of the correct sign in the relay feedback tests, then the undesirable pairing can be eliminated in the process. Unfortunately, it can only be used to eliminate *undesirable* pairings, but not for finding the *best* pairing.

6.4.4 Summary of Procedure

From the properties discussed, an effective procedure for multivariable autotuning is proposed. It depends on the extent of the process information available. Process information is classified into "required" and "helpful". The "signs" of the diagonal elements are the required input data. As mentioned earlier, as the control hardware is already installed, it is nearly impossible not to know the sign of the steady state gain (*e.g.* not to know whether the controller is direct or reverse acting for the diagonal elements). The relative speed of diagonal transfer function, *e.g.* s_i in Equation 6.32, is helpful, since knowing this can lead to faster convergence for systems with very different loop speeds. With the process information available, the automatic tuning procedure for an $n \times n$ multivariable system becomes (Figure 6.16):

(0) rank the loop speed from fast to slow into 1, 2,..., n
(1a) perform the relay feedback test on loop 1 (the relay is switched according to the sign of g_{ii})
(1b) design k_1 using the modified Ziegler–Nichols method (Equations 6.19 and 6.20) and put k_1 into automatic
(2a) perform the relay feedback test on loop 2
(2b) design k_2 using the modified Ziegler–Nichols method and put k_2 into automatic
$\vdots$
(na) perform the relay feedback test on loop n
(nb) design k_n using the modified Ziegler–Nichols method and put k_n into automatic.

This procedure is repeated (back to step 1) until the controller parameters converge. Typically, a total of $n + (n-1)$ identification-design steps will suffice, as shown in the next section. Figure 6.16 shows the flow chart of the MIMO autotuning procedure.

6.5 Applications

Two nonlinear distillation columns and a 3×3 linear example are used to illustrate the MIMO autotuning procedure. For the nonlinear examples, one is a moderate-purity column [31] and the other is a high-purity column [32]. The 3×3 linear system is a transfer function matrix for a distillation column (T4 column [8]).

6.5.1 Moderate-purity Column

This is a 20-tray column studied by Shen and Yu [31]. The product specifications are 98% and 2% of the light component on the top and bottoms of the column. The relative volatility is 2.26 with a reflux ratio 1.76. Table 5.1 summarizes the steady state operating conditions. The control objective is to maintain the top and bottoms compositions ($x_D = 0.98$ and $x_B = 0.02$) by changing the reflux flow rate R and vapor boil-up V. This is the conventional $R-V$ control structure (Figure 5.12). In the nonlinear simulation, the following assumptions are made:

124 Autotuning of PID Controllers

1. equal molar overflow
2. 100% tray efficiency
3. saturated liquid feed
4. total condenser and partial reboiler
5. perfect level control.

According to the autotuning procedure in Figure 6.16, relay feedback tests are performed on the $R-V$ controlled distillation column. The relay heights h of 3% are used in R and V. The $x_D - R$ loop is tuned first and the controller k_1 is designed (Equations 6.19 and 6.20) followed by the second relay feedback test on the $x_B - V$ loop (Figure 6.17). After the bottoms controller k_2 is designed and put on automatic, the third relay feedback experiment is performed and the controller parameters for k_1 are finalized. The resultant controller parameters are $K_{c1} = 245.34$, $\tau_{I1} = 97.1$ and $K_{c2} = -157.4$, $\tau_{I2} = 47.8$. The results of the first ($t = 0$–200 min) and the third ($t = 400$–600 min) relay feedback tests clearly show that the controllers k_1 designed from these two tests are almost the same ($K_{c1} = 223.25$ and 245.36 and $\tau_{I1} = 89.0$ and 97.1). This means, in fact, two relay feedback tests are sufficient for this column. A +5% step flow rate change is introduced at $t = 800$ min and a +5% feed composition change is also made at $t = 1300$ min. Figure 6.17 shows that good load rejections are achieved with the tuning constants from the autotuning procedure. It should be emphasized that the good load responses are achieved with little engineering effort (no transfer func-tion is fitted, no frequency-domain plot is generated). The only knowledge assumed is that one has to know the "sign" between x_D and R is positive and the "sign" between x_B and V is negative. Actually, this information is not applied, since the outputs and inputs are paired correctly.

6.5.2 High-purity Column

The high-purity column is a C_3 splitter [32] which separates propane and propylene in a 190-tray column. The relative volatility ranges from 1.12 to 1.24 and the product specifications are 99.66% and 0.02% light component (propylene) at the top and bottoms of the column respectively. This is a very high purity column with difficult separation. In terms of control, this is a highly nonlinear system [33]. The $D-B$ control structure is considered for dual-composition control (Figure 6.18). The steady- state operating conditions are given in Chang and Yu [20]. The autotuning procedure is applied to the C_3 splitter. The $x_D - D$ loop is tuned first and the autotuning is terminated in three relay feedback experiments. Again, the controller parameters for the $x_D - D$ loop are expected to be almost the same from the first and the third relay feedback tests (Figure 6.19). Table 6.4 shows the corresponding tuning constants from the autotuning steps. A 5% feed composition change is made at $t = 1000$ min (Figure 6.19) and simulation results show that good load responses are obtained. Chang and Yu [20] also studied the modeling, tuning and robustness aspects of the C_3 splitter. Since this column is highly nonlinear and a non-conventional control structure is employed (resulting in a

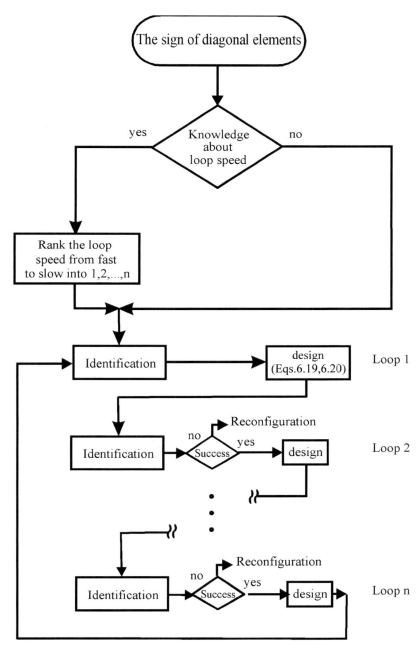

Figure 6.16. MIMO autotuning procedure

126 Autotuning of PID Controllers

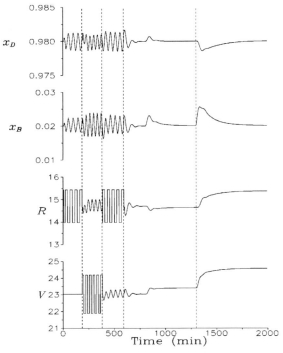

Figure 6.17. Automatic tuning and load responses for a 5% step change in feed composition for the Shen and Yu column

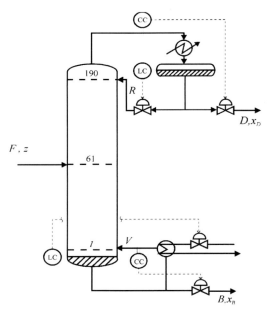

Figure 6.18. D–B control structure for C_3 splitter

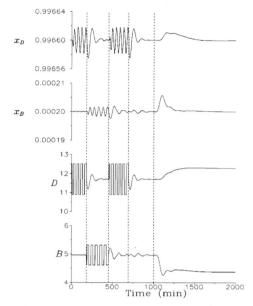

Figure 6.19. Automatic tuning and load responses for a 5% step change in feed composition for the C_3 splitter

transfer function with a $1/s$ term), a great deal of effort is spent finding the transfer function matrix [32]. Chang and Yu [20] use the stepping technique (see Luyben [26] p.444) to find frequency responses for the $R-V$ structure followed by a transformation from $R-V$ to $D-B$ structure. Once the transfer function matrix is obtained, a modified version of singular value tuning (SVT) [34] is used to find the tuning constants for the PI controllers [20]. Table 6.4 gives the autotuning constants from the SVT.

Comparisons are made between the autotuning approach and the SVT tuned control system. Figure 6.20 shows the load responses for ±30% feed composition changes. Similar results can be seen for feed flow rate disturbances.

The results show that better load responses are achieved using the proposed autotuning procedure. More importantly, good performance is achieved with very little engineering effort. The multivariable autotuner is activated at $t = 0$ and the tuning is completed at $t = 700$ min following three relay feedback tests.

Table 6.4. Controller parameters for a C_3 splitter using the proposed autotuning procedure and SVT

Tuning method	Step 1 $x_D - D$ loop (K_c/τ_I)	Step 2 $x_B - B$ loop (K_c/τ_I)	Step 3 $x_D - D$ loop (K_c/τ_I)
MIMO autotuner	150.42/93.42	42.28/108.32	150.08/91.68
SVT method	45.64/79.35	7.91/150.0	

128 Autotuning of PID Controllers

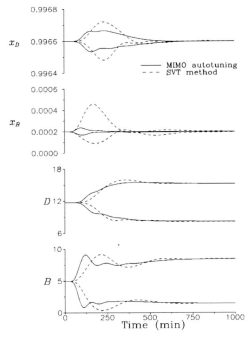

Figure 6.20. Load responses for ±30% feed composition change with different design methods

6.5.3 T4 Column

This autotuning procedure can be extended to an $n \times n$ multivariable system in a straightforward manner. The following 3×3 literature example [8] illustrates this extension. The transfer function matrix of the T4 column is

$$\begin{pmatrix} y_1 \\ y_2 \\ y_3 \end{pmatrix} = \begin{pmatrix} \dfrac{-2.986e^{-0.71s}}{66.7s+1} & \dfrac{-5.24e^{-60s}}{400s+1} & \dfrac{-5.984e^{-2.24s}}{14.29s+1} \\ \dfrac{0.0204e^{-0.59s}}{(7.14s+1)^2} & \dfrac{-0.33e^{-0.68s}}{(2.38s+1)^2} & \dfrac{2.38e^{-0.42s}}{(1.43s+1)^2} \\ \dfrac{0.374e^{-7.75s}}{22.22s+1} & \dfrac{-11.3e^{-3.79s}}{(21.74s+1)^2} & \dfrac{-9.811e^{-1.59s}}{11.36s+1} \end{pmatrix} \begin{pmatrix} u_1 \\ u_2 \\ u_3 \end{pmatrix}$$

Following the autotuning procedure, the loop speed is ranked from fast to slow as: loop 1, loop 2, loop 3. The autotuning is carried out according to Figure 6.16 and the resultant controller parameters are $K_{c1} = -2.51$, $K_{c2} = -6.34$, $K_{c3} = -0.23$ and $\tau_{I1} = 16.2$, $\tau_{I2} = 12.4$, $\tau_{I3} = 12.6$. Figure 6.21 shows that the tuning procedure is completed in the first 180 min and an SP change is made at $t = 300$ min.

Notice that, except for K_{c1}, the controller parameters from five relay feedback tests are essentially the same as the resultant parameters ($K_{c1} = -3.24$, $K_{c2} = -6.05$, $K_{c3} = -0.28$ and $\tau_{I1} = 12.05$, $\tau_{I2} = 12.17$, $\tau_{I3} = 12.37$).

The results (Figure 6.21) show that the autotuning procedure gives reasonable servo responses. Actually, the modified Ziegler–Nichols method shows better SP responses than the BLT method. Again, the results show that the proposed autotuning procedure achieves good performance with very little engineering effort for an HO (3×3) system.

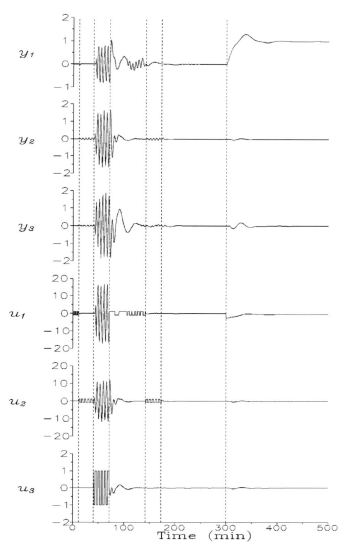

Figure 6.21. Automatic tuning and SP change in loop 1 for T4 column

6.6 Conclusion

In this chapter we have seen the multivariable version of the Åström–Hägglund autotuner. It is based on the concept of sequential identification-design. The consistent nature (satisfying consistency relations) of the sequential identification is discussed and the advantages are shown. Once the ultimate properties become available, a modified Ziegler–Nichols tuning is applied to find the settings sequentially. The convergent nature of the multivariable autotuner is also conjectured. It is important to recognize that this relatively simple autotuner works and works well for very difficult processes: recycle chemical process (Chapter 11), reactive distillation [34], Tennessee Eastman process (Chapter 8), *etc*. It is used routinely in controller tuning or complex processes, and we suggest you try it on your systems.

6.7 References

1. Koivo HN, Pohjolainen S. Tuning of multivariable PI controller for unknown systems with input delay. Automatica 1985;21:81.

2. Cao R, McAvoy TJ. Evaluation of pattern recognition adaptive PID controller. Automatica 1990;26:797.

3. Hsu L, Chan M, Bhaya A. Automated synthesis of decentralized tuning regulators for systems with measurable DC gain. Automatica 1992;28:185.

4. Åström KJ, Hägglund T. Automatic tuning of simple regulators with specifications on phase and amplitude mMargins. Automatic 1984;20:645.

5. Åström KJ, Hang CC, Persson P, Ho WK. Towards intelligent PID control. Automatica 1992;28:1.

6. Schei TS. A method for closed loop automatic tuning of PID controllers. Automatica 1992;28:587.

7. Chiu MS, Arkun Y. A methodology for sequential design of robust decentralized control systems. Automatica 1992;28:997.

8. Luyben WL. Simple method for tuning SISO controllers in multivariable systems. Ind. Eng. Chem. Process Des. Dev. 1986;25:654.

9. Skogestad S, Landström P. μ-optimal LV-control of distillation columns. Comput. Chem. Eng. 1990;14:401.

10. Mayne DQ. The design of linear multivariable systems. Automatica 1973;9:201.

11. Mayne DQ. Sequential design of linear multivariable systems. Proc. IEE Part D 1979;126:568.

12. Bernstein DS. Sequential design of decentralized dynamic compensators using the optimal projection equations. Int. J. Control 1987;46:1569.

13. O'Reilly J, Leithead WE. Multivariable control by individual channel design. Int. J. Control 1991;54:1.

14. Leithead WE, O'Reilly J. Performance issues in the individual channel design of 2-input 2-output systems. Part 1. Structural issues. Int. J. Control 1991;54:47.

15. Bhalodia M, Weber TW. Feedback control of a two-input, two-output interacting process. Ind. Eng. Chem. Process Des. Dev. 1979;18:599.

16. Grosdidier P, Morari M, Holt RB. Closed-loop properties from steady-state gain information. Ind. Eng. Chem. Process Des. Dev. 1985;24:221.

17. Yu CC, Fan MKH. Decentralized integral controllability and D-stability. Chem. Eng. Sci. 1990;45:3299.

18. Bristol EH. New measure of interaction for multivariable process control. IEEE Trans. Automat. Control 1966;AC-11:133.

19. Luyben WL. Derivation of transfer functions for highly nonlinear distillation columns. Ind. Eng. Chem. Res. 1987;26:2490.

20. Chang DM, Yu CC. The distillate-bottoms control of distillation columns: Modeling, tuning and robustness issues. J. Chin. Inst. Chem. Eng. 1992;23:344.

21. Luyben WL. Sensitivity of distillation relative gain arrays to steady-state gains. Ind. Eng. Chem. Res. 1987;26:2076.

22. Häggblom KE, Waller KV. Transformations and consistency relations of distillation control structures. AIChE J. 1988;34:1634.

23. Häggblom KE, Waller KV. Control structure, consistency, and transformations. Practical distillation control. Luyben. WL. ed. New York: Van Nostrand Reinhold; 1992.

24. Ziegler JG, Nichols NB. Optimum settings for automatic controllers. Trans. ASME 1942;12:759.

25. Seborg DE, Edgar TF, Mellichamp DA. Process dynamics and control. 2nd ed. New York: Wiley; 2004.

26. Luyben WL. Process modeling, simulation and control for chemical engineers. 2nd ed. New York: McGraw-Hill; 1990.

27. Tan LY, Weber TW. Controller tuning of a third-order process under proportional–integral control. Ind. Eng. Chem. Process Des. Dev. 1985;24:155.

28. Marino-Galarraga M, McAvoy TJ, Marlin TE. Short-cut operability analysis. 2. estimation of f_i detuning parameter for classical control systems. Ind. Eng. Chem. Res. 1987;26:511.

29. Rice JR. Numerical methods, software, and analysis. New York: McGraw-Hill; 1983.

30. Shen SH, Yu CC. Use of relay feedback test for automatic tuning of multivariable systems. AIChE J. 1994;40:627.

31. Shen SH, Yu CC. Indirect feedforward control: Multivariable systems. Chem. Eng. Sci. 1992;47:3085.
32. Papastathopoulou HS, Luyben WL. Tuning controllers on distillation columns with the distillate-bottoms structure. Ind. Eng. Chem. Res. 1990;29:1859.
33. Fuentes C, Luyben WL. Control of high-purity distillation columns. Ind. Eng. Chem. Process Des. Dev. 1983;22:361.
34. Chiang TP, Luyben WL. Comparison of the dynamic performances of three heat-integrated distillation configurations. Ind. Eng. Chem. Res. 1988;27:99.
35. Huang SG, Kuo CL, Hung SB, Chen YW, Yu CC. Temperature control of heterogeneous reactive distillation: Butyl propionate and butyl acetate esterification. AIChE J., 2004;50:2203.

Appendix

Consider a 2×2 system under decentralized PI control. Assume that:

1. g_{ij} are rational, strictly proper transfer functions with no RHP pole.
2. The closed-loop system is stable along the tuning sequence (loop 1 tuned first, followed by loop 2 then back to loop 1).
3. No pole–zero cancellation occurs in $g_{ii,CL}$.

If $\kappa(0)>1$ (or $\lambda_{11}<0$) and loop 1 is tuned first, then $g_{11,CL}(s)$ has at least one RHP pole.

Proof: Since loop 1 is tuned first (with loop 2 on manual) and the system is stable, this means $1+g_{11}(s)k_1(s)=0$ does not have an RHP zero or, equivalently, h_1 does not have an RHP pole.

Let g_{22} take the form

$$g_{22}(s) = K_{p22} \frac{N(s)}{D(s)} \tag{A1}$$

where $N(s)$ and $D(s)$ are numerator and denominator polynomials with deg$(D(s))>$deg$(N(s))$ and (from the strictly proper assumption) $D(0)=N(0)=1$. The second controller k_2 is designed according to

$$g_{22,CL} = g_{22}(s)(1-\kappa(s)h_1(s)) \tag{A2}$$

Since $\kappa(0)>1$ and $h_1(0)=1$, it then becomes obvious that

$$g_{22,CL}(0) = g_{22}(0)(1-\kappa(0)) \tag{A3}$$

This indicates that the sign of $g_{22,CL}$ is different from that of g_{22}. Since $g_{22,CL}$ does not have an RHP pole (the poles of $g_{22,CL}$ are the poles of g_{22}, g_{12}, g_{21} and

h_1) and the closed-loop system $1+g_{22,CL}k_2 = 0$ is stable, the controller gain K_{c2} has the same sign as that of $g_{22,CL}$:

$$g_{22,CL}(0)K_{c2} > 0 \qquad (A4)$$

or

$$g_{22}(0)K_{c2} < 0 \qquad (\text{or } K_{p22}K_{c2} < 0) \qquad (A5)$$

Now go back to loop 1 to design k_1 for $g_{11,CL}$. Notice that the poles of $g_{11,CL}$ are the poles of g_{11}, g_{12}, g_{21} and h_2. Let us consider the poles of h_2. The zeros of the closed-loop characteristic equation become

$$1 + g_{22}K_{c2}\frac{\tau_I s + 1}{\tau_I s} = 0 \qquad (A6)$$

$$1 + \frac{K_{p22}K_{c2}(\tau_I s + 1)}{\tau_I s}\frac{N(s)}{D(s)} = 0 \qquad (A7)$$

$$\tau_I s D(s) + K_{p22}K_{c2}\tau_I s N(s) + K_{p22}K_{c2} N(s) = 0 \qquad (A8)$$

Since $D(0)=1$ and g_{22} is stable and strictly proper, the coefficient of the highest degree is positive. From $N(0)=1$, the constant term $K_{p22}K_{c2}$ is negative. It is obvious that the closed-loop characteristic equation has at least one RHP zero. This implies h_2 or $g_{11,CL}$ has at least one RHP pole. ∎

7
Load Disturbance

Since frequent and-large load changes are often encountered in industrial processes, any identification procedure should be able to, at least, detect load changes, more positively, to find a quality process model under load disturbance. In this chapter, the relay feedback system is enhanced by taking the effect of the load into account. In other words, instead of analyzing or validating the plant data afterward, the load effect is *compensated* for during the plant tests.

7.1 Problems

Disturbance rejection is the major consideration in chemical process control. In terms of control, slow chemical processes have to overcome frequent and progressive types of load change. Facing frequent load changes, any reliable identification methodology should be as insensitive to load change as possible. Therefore, sensitivity with respect to load changes is an important criterion in evaluating identification techniques.

7.1.1 Step Change versus Continuous Cycling

Two popular identification methods, step test and relay feedback test, are compared. Consider a second-order linear system, the process $G_2(s)$ and load $G_L(s)$ transfer functions are

$$G_2(s) = \frac{e^{-2s}}{(10s+1)(s+1)} \tag{7.1}$$

$$G_L(s) = \frac{e^{-s}}{(10s+1)} \tag{7.2}$$

136 Autotuning of PID Controllers

First, a step test is introduced and the step response is shown in Figure 7.1. From the system input u and output y, the reaction curve method [1,2] is used to find an SOPDT model. The result gives

$$\hat{G}_2(s) = \frac{e^{-2s}}{(9.4s+1)(0.86s+1)} \qquad (7.3)$$

From the identified model $\hat{G}_2(s)$, the ultimate gain K_u and ultimate frequency ω_u are $K_u = 6.749$ and $\omega_u = 0.6237$. This corresponds to -4.5% error in K_u and 3.95% error in ω_u. Actually, a fairly good model is found from the process reaction curve. However, if a step load change ($L = 0.5$) comes in at $t = 10$, then the step response is distorted, as shown in Figure 7.1 (solid line). Again, Smith's method is employed to find the transfer function. It becomes

$$\hat{G}_2(s) = \frac{1.5e^{-2s}}{(8s+1)(2.3s+1)} \qquad (7.4)$$

From the identified model ($\hat{G}_2(s)$ in Equation 7.4), it is quite clear that the quality of the model deteriorates significantly. This corresponds to -42.7% and -18.3% errors in K_u and ω_u respectively. Furthermore, the deviation can become even greater if the magnitude of load change increases.

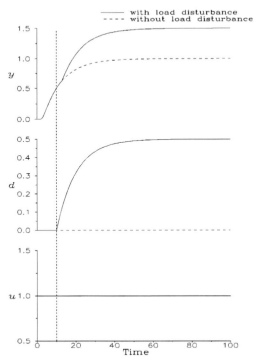

Figure 7.1. Step test for $G_2(s)$ with and without load disturbance

An alternative is the relay feedback test, where an ideal relay is placed in the feedback loop to generate a sustained oscillation. From system responses, the important information, ultimate gain K_u and ultimate frequency ω_u, can be found. This gives

$$K_u = \frac{4h}{\pi a} \quad (7.5)$$

$$\omega_u = \frac{2\pi}{P_u} \quad (7.6)$$

where h is the height of the relay and a is the *amplitude* of the output response. Figure 7.2 shows the input and output responses and K_u and ω_u can be calculated from Equations 7.5 and 7.6. This gives $K_u = 6.55$ and $\omega_u = 0.5978$, which corresponds to -7.36% and -0.11% errors in K_u and ω_u respectively. Again, a load change with $L = 0.5$ is introduced at $t = 10$ min (solid line in Figure 7.2). From system responses, two observations become apparent. First, the system output is asymmetric (y is asymmetric with respect to the SP). The second observation is that the offset in y is not equal to the magnitude of the load effect d (Figure 7.2). For this asymmetric oscillation, the amplitude of the oscillation a is taken as the average of the oscillation:

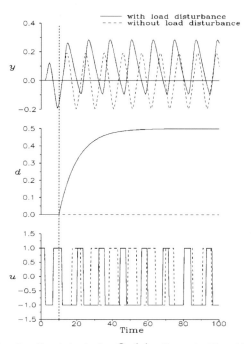

Figure 7.2. Relay feedback tests for $G_2(s)$ with and without load disturbance

$$a = \frac{y^{max} - y^{min}}{2} \tag{7.7}$$

where y^{max} and y^{min} stand for the maximum and minimum amplitudes of the process output respectively. Despite the presence of the load disturbance, the resulting estimates of K_u and ω_u are still quite reliable and the corresponding errors in K_u and ω_u are −7.14% and −13.1% respectively.

The results clearly indicate that the continuous cycling using an ideal relay feedback is more robust with respect to load change than the method of step test. This can be understood since the limit of stability is invariant under bounded load changes. Therefore, in theory, K_u and ω_u cannot be affected by load disturbance.

7.1.2 Effect of Load Change on Relay Feedback Test

Despite the fact that the relay feedback test is less sensitive to load disturbances, the estimates of K_u and ω_u (in particular) also deteriorate slightly for a moderate load change ($L = 0.5$). Unfortunately, the errors in the estimates of K_u and ω_u grow exponentially as the magnitude of the load change increases. Let us take three typical linear transfer functions illustrating the effect of load disturbance.

Consider an ideal relay feedback system with an external load variable L (Figure 7.3A). Under load disturbance, an ideal relay feedback test results in an asymmetrical oscillation, as shown in Figure 7.3B. In addition to the SOPDT system ($G_2(s)$ in Equation 7.1), a first-order and a third-order plus dead time system are considered.

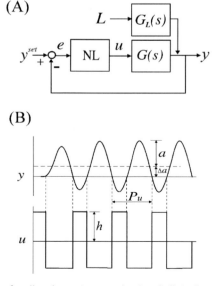

Figure 7.3. Ideal relay feedback system under load disturbance: (A) block diagram and (B) system input and output responses

$$G_1(s) = \frac{e^{-2s}}{(10s+1)} \quad (7.8)$$

and

$$G_3(s) = \frac{e^{-2s}}{(20s+1)(10s+1)(s+1)} \quad (7.9)$$

with a load transfer function $G_L(s)$ (Equation 7.2). Ideal relay feedback tests with $h = 1$ are performed on these three examples for a wide range of load changes. K_u and ω_u (or P_u) are computed from system input and output responses and the results show that estimated errors, ω_u in particular, grow rapidly as the magnitude of the load increases (Figure 7.4). Furthermore, the estimated K_u deteriorates as the order of the system increases. Since load disturbance is uncontrollable in an operating environment, remedial action has to be taken to ensure the quality of the identified model.

7.2 Analyses

7.2.1 Causes of Errors

An ideal relay feedback test under a step load change gives asymmetric output responses (y in Figure 7.3) and, consequently, an imbalance in half periods results. This asymmetry and the imbalance lead to errors in the estimates of K_u and ω_u.

Unlike a simple relay feedback system ($L = 0$ in Figure 7.3A), the input to the nonlinear element consists of two elements: a symmetric oscillation and a step input. This type of problem was known to the control community as the dual input describing function (DIDF) as early as the 1950s [3]. In order to describe the characteristics of the nonlinear element (an ideal relay), the relationship between the input and output signals of the nonlinear element can be separated into two parts: one is the oscillatory part (the gain of the sinusoidal wave to the output of the nonlinear element) and the other is the static part (the gain of the biased signal to the output of a nonlinear element). The static part can be described by the equivalent gain [4]:

$$N_\gamma = \frac{\bar{u}}{\bar{e}} \quad (7.10)$$

where $\bar{u}$ and $\bar{e}$ are respectively the averaged input and output of the nonlinear element, which can be found by integrating the system response. For example:

$$\bar{u} = \int_t^{P_u+t} u(t)\, dt \quad (7.11)$$

$$\bar{e} = \int_t^{P_u+t} e(t)\, dt \quad (7.12)$$

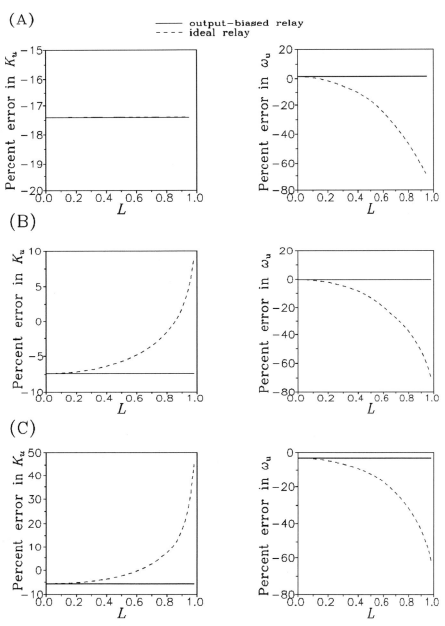

Figure 7.4. Percentage error in K_u and ω_u under load changes for ideal and output-biased relays: (A) first-order model (G_1), (B) second-order model (G_2), and (C) third-order model (G_3)

The oscillatory part is characterized by the familiar describing function $N(a)$. The following conditions should be satisfied for the existence of a sustained oscillation [5]:

$$\bar{e}(1+N_\gamma G(0)) = -G_L(0)L \tag{7.13}$$

$$1+NG(j\omega_u) = 0 \tag{7.14}$$

Equations 7.13 and 7.14 give constraint at the low ($\omega=0$) and high ($\omega=\omega_u$) frequencies. Furthermore, Equation 7.13 relates the bias in the output $\bar{e}$ to the external load disturbance $G_L L$.

The describing function analysis shows the difference between a simple relay and a relay feedback under load change. Consider a relay feedback system (Figure 7.3). The output $u(t)$ of the nonlinear element can be expressed in terms of Fourier series:

$$u(t) = A_0 + \sum_{n=1}^{\infty}(A_n \cos n\omega t + B_n \sin n\omega t) \tag{7.15}$$

where

$$\begin{aligned} A_0 &= \frac{1}{2\pi}\int_0^{2\pi} u(t)\,d\omega t \\ &= -\frac{2h}{\pi}\sin^{-1}\left(\frac{\Delta a}{a}\right) \end{aligned} \tag{7.16}$$

$$\begin{aligned} A_n &= \frac{1}{\pi}\int_0^{2\pi} u(t)\cos n\omega t\,d\omega t \\ &= \frac{4h}{n\pi}\sin\left[n\sin^{-1}\left(\frac{\Delta a}{a}\right)\right], \quad n=2,4,6,\cdots \end{aligned} \tag{7.17}$$

$$\begin{aligned} B_n &= \frac{1}{\pi}\int_0^{2\pi} u(t)\sin n\omega t\,d\omega t \\ &= \frac{4h}{n\pi}\cos\left[n\sin^{-1}\left(\frac{\Delta a}{a}\right)\right], \quad n=1,3,5,\cdots \end{aligned} \tag{7.18}$$

The factor A_0 describes the imbalance in the half periods. Furthermore, in addition to the well-known terms B_1, B_3, B_5, $\cdots$ from a single-input system ($L=0$), the dual-input system ($L \neq 0$) gives the following non-zero terms: A_2, A_4, A_6, $\cdots$. It is quite clear that the term describing the imbalance, A_0, and the additional terms, A_2, A_4, A_6, etc. degrade the principal harmonic approximation and, subsequently, lead to erroneous estimates of K_u and ω_u. Furthermore, if Δa approaches zero, the A_0, A_2, A_4, $\cdots$ terms disappear. Notice that the asymmetric output response can also be observed from a relay feedback under a disturbance-free condition (Figure 7.5). Figure 7.5A shows an output-biased relay with a bias

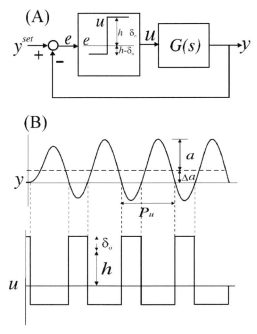

Figure 7.5. Output-biased relay feedback system: (A) block diagram and (B) system input and output responses

value δ_0. The system response indicates an asymmetrical oscillation with respect to the SP ($y = 0$) and the imbalance in half periods is also observed (Figure 7.3). It then becomes obvious that we can utilize the asymmetry (δa_0 in Figure 7.5B) of the output-biased relay to cancel out the asymmetry (Δa in Figure 7.3B) generated from a load change.

7.2.2 Output-biased Relay Feedback System

Before getting into the detail of restoring symmetry, the frequency response of the output-biased relay is analyzed [6]. Consider an output-biased relay feedback system (Figure 7.5): the output of the nonlinear element u (Figure 7.5) can be expressed in terms of Fourier coefficients, *i.e.*

$$A_0 = \frac{1}{2\pi} \int_0^{2\pi} u(t)\, d\omega t$$
$$= -\frac{2h}{\pi} \sin^{-1}\left(\frac{\Delta a_0}{a}\right) - \delta_0 \qquad (7.19)$$

$$A_n = \frac{1}{\pi} \int_0^{2\pi} u(t) \cos n\omega t \, d\omega t$$
$$= \frac{4h}{n\pi} \sin\left[n \sin^{-1}\left(\frac{\Delta a_0}{a}\right)\right], \quad n = 2, 4, 6, \cdots \quad (7.20)$$

$$B_n = \frac{1}{\pi} \int_0^{2\pi} u(t) \sin n\omega t \, d\omega t$$
$$= \frac{4h}{n\pi} \cos\left[n \sin^{-1}\left(\frac{\Delta a_0}{a}\right)\right], \quad n = 1, 3, 5, \cdots \quad (7.21)$$

Equations 7.19–7.21 show all non-zero terms in the Fourier expansion. Except for the biased term (Equation 7.19), Fourier coefficients describing the oscillatory part (Equations 7.20 and 7.21) are exactly the same as those from load disturbance (Equations 7.17 and 7.18).

Let us first consider an ideal relay feedback system. If a load disturbance is introduced, then the output oscillation is biased with a value of Δa (Figure 7.3). If the relay block is switched to an output-biased relay (*e.g.* Figure 7.5A), with an appropriate adjustment of δ_0, it is possible to eliminate the asymmetry in the output oscillation. That implies that, under a static load change, we can have an output-biased relay feedback system with $\Delta a_0 = 0$. Under this circumstance ($\Delta a_0 = 0$), Equation 7.24 becomes

$$A_n = 0, \quad n = 2, 4, 6, \cdots \quad (7.22)$$

$$B_n = \frac{4h}{n\pi}, \quad n = 1, 3, 5, \cdots \quad (7.23)$$

and the principal harmonic approximation gives the following describing function:

$$N(a) = \frac{4h}{\pi a} \quad (7.24)$$

It should be emphasized that Equation 7.24 is exactly the same as the describing function for the disturbance-free case. More importantly, performance degradation (in the estimate of K_u and ω_u) can be eliminated. The output-biased relay feedback can retain the quality of the estimates for a wide range of load changes. But, the correct δ_0 comes from a trial-and-error procedure [7].

In an operating condition, it is not practical to have an on-line trial-and-error procedure, since this can prolong the duration time for plant test. Therefore, it is desirable to devise a procedure with which the biased value δ_0 can be found efficiently.

7.2.3 Derivation of Bias Value δ_0

7.2.3.1 Effect of Load Disturbance

Before finding an appropriate δ_0 to overcome load effect, one has to analyze the relationship between the output bias Δa and the load effect from the available process information. From the literature, one possibility is the static relationship describing the existence of a sustained oscillation (Equation 7.13). That is:

$$\bar{e}\left(1+\frac{\bar{u}}{\bar{e}}K_p\right) = -K_L L \tag{7.25}$$

where $\bar{e}$ and $\bar{u}$ can be found from integrating system output and input responses and K_p and K_L stand for steady state gains for the process and load transfer functions respectively. Unfortunately, Equation 7.25 gives two unknowns K_p and $K_L L$ that cannot be solved explicitly. Furthermore, Δa is not involved in the equation.

A new relationship describing Δa and load effect $K_L L$ is derived. Let us analyze two extreme conditions for the existence of a sustained oscillation. The asymmetry in the output Δa arises from the A_0 term (Equation 7.16) according to the Fourier expansion. Therefore, the two cases are classified as the lower and upper bounds of Δa under the condition for the existence of a limit cycle. Consider the feedback loop in Figure 7.3.

(1) *Lower bound* $\Delta a = 0$

The lower bound for Δa is quite obvious, *i.e.* $\Delta a = 0$. The corresponding load effect is

$$K_L L = 0 \tag{7.26}$$

This can be understood from the typical symmetric oscillation obtained from an ideal relay feedback when load disturbance does not exist (symmetric input–output responses is an indication of zero load effect). This can also be understood by analyzing Fourier coefficients (Equations 7.16 and 7.18). If $\Delta a = 0$, then the bias term A_0 (Equation 7.16) disappears and this implies that there is no asymptotically constant load change.

(2) *Upper bound* $\Delta a \to a$

Another extreme is when Δa approaches the amplitude of the output oscillation (Figure 7.3B). It should be noted that we cannot have a Δa which exceeds the amplitude a and still has a sustained oscillation. Taking the limit, the bias term of the Fourier coefficient (Equation 7.16) becomes

$$A_0 \to -h \tag{7.27}$$

Furthermore, the rest of the non-zero Fourier coefficients, A_2, A_4, $\cdots$ and B_1, B_3, $\cdots$, are approaching zero (Equations 7.17 and 7.18). Equation 7.27 provides another perspective to the limiting condition: it characterizes the manipulated input $u(t)$. Therefore, instead of correlating Δa to the maximum allowable load change $K_L L$, the relationship between the manipulated input and largest load change for the existence of a limit cycle is established. A steady state analysis from the block diagram (Figure 7.3) indicates that, under this circumstance, the maximum allowable load change is in balance with the achievable capacity of the manipulated input. That is:

$$K_L L + K_p \cdot (-h) = 0 \tag{7.28}$$

or

$$K_L L = K_p h \tag{7.29}$$

Let us take the first-order example G_1 to illustrate this. Consider the case when $h = 1$ and $L = 0.999999$ ($K_L L < K_p h$); the input and output responses are shown in Figure 7.6. The results indicate that the magnitude of the manipulated input $u(t)$ approaches $-h$ while the output y stays fairly close to the SP ($y = 0$). Furthermore, if $K_L L$ exceeds unity, i.e. $K_L L > K_p h$, then the ideal relay feedback system fails to generate a sustained oscillation.

After deriving the relationship between the load effect $K_L L$ and corresponding process variables (e.g. K_p and h) at these two extremes, a new result can readily be formulated. For linear processes, the linear interpolation can be utilized to correlate the asymmetry Δa to the load effect $K_L L$. From Equations 7.26 and 7.29, the slope describing Δa and $K_L L$ is

$$\frac{a-0}{K_p h - 0} = \frac{a}{K_p h} \tag{7.30}$$

Therefore, the linear relationship relating the asymmetry Δa to the load effect becomes

$$\Delta a = \frac{a}{K_p h} K_L L \tag{7.31}$$

or

$$\frac{\Delta a}{a} = \left(\frac{1}{K_p h}\right) K_L L \tag{7.32}$$

The result gives a simple description between the load effect and relevant process variables. Again, the three linear examples (G_1, G_2 and G_3) are used to validate Equation 7.32. Figure 7.7 shows that for the three typical process transfer functions with a load transfer function (G_L in Equation 7.2), the asymmetry $\Delta a / a$ is indeed is a linear function of the load effect ($K_L L$). This is the main result (Equation 7.32) in this chapter.

Furthermore, with this new equation, we are able to back-calculate the load effect and the steady state gain. Substituting $K_L L$ of Equation 7.32 into Equation 7.25, the steady state gain can be solved directly. That gives:

$$K_p = \frac{-\overline{e}}{\left(\frac{\Delta a}{a}\right) h + \overline{u}} \tag{7.33}$$

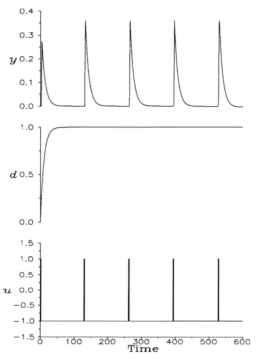

Figure 7.6. Input and output responses for an ideal relay feedback system with the magnitude of load change approaching the upper bound ($K_L L \rightarrow K_p h$) for G_1

Once K_p is available, the load effect can be solved from Equation 7.32.

$$K_L L = \left(\frac{\Delta a}{a}\right) K_p h \tag{7.34}$$

Therefore, Equations 7.33 and 7.34 can be utilized to find additional process information. Since the objective of this work is to restore a symmetric output response, the use of these two equations will not be elaborated further.

7.2.3.2 Opposite Effect from Output-biased Relay

Similarly, the counterpart of the linear relationship can be derived for the output-biased relay feedback system. Again, from the Fourier analysis, the two extreme conditions for the existence of a limit cycle can readily be derived. The lower bound corresponds to

$$\delta_0 = 0 \Rightarrow \Delta a_0 = 0 \tag{7.35}$$

and the upper bound is

$$\delta_0 \rightarrow 0 \Rightarrow \Delta a_0 \rightarrow 0 \tag{7.36}$$

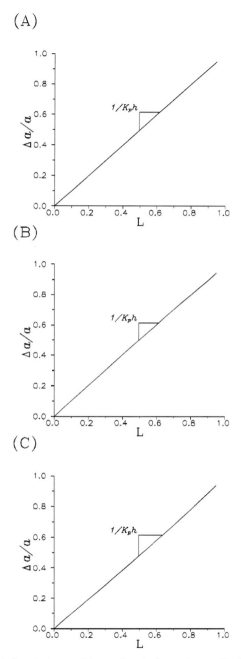

Figure 7.7. Linear relation between bias value $\Delta a/a$ and magnitude of load L under ideal relay feedback for: (A) first-order (G_1), (B) second-order (G_2) and (C) third-order (G_3) systems

Therefore, the linear relationship for the output-biased relay becomes

$$\frac{\Delta a_0}{a} = \left(\frac{1}{h}\right)\delta_0 \qquad (7.37)$$

Actually, Equation 7.37 is the dual of Equation 7.32, which describes the relationship between the asymmetry in the output response Δa_0 and the biased value δ_0 in the relay.

Since the load change introduces the asymmetry Δa in y, the way to restore the symmetric output response is to switch the relay to an output-biased relay such that

$$\Delta a_o = -\Delta a \qquad (7.38)$$

Therefore, the bias value δ_0 can be found immediately (Equations 7.37 and 7.38).

$$\delta_0 = -\left(\frac{\Delta a}{a}\right)h \qquad (7.39)$$

This important result gives the bias value δ_0 for a better estimate of K_u and ω_u throughout the relay feedback test. Without Equation 7.39, one has to go through a trial-and-error procedure to find the correct value of δ_0 [7]. More importantly, all the variables (Δa, a and h) for computing δ_0 can be read off directly from the input–output response.

7.3 Summary of Procedure

Consider an ideal relay feedback system (Figure 7.3A). If a load disturbance comes into the system during the plant test, an asymmetric oscillation results. If information about the magnitude of disturbance and steady state gain is needed, Equations 7.33 and 7.34 can be utilized to solve for K_p and $K_L L$ provided with $\bar{u}$ and $\bar{y}$. Otherwise, the experiment proceeds with an output-biased relay using the bias value computed from Equation 7.39. Therefore, the procedure can be summarized as follows:

(1) Perform an ideal relay feedback test (Figure 7.3).

    (1a) If the limit cycle is symmetric, then calculate K_u and ω_u

$$K_u = \frac{4h}{\pi a}$$

$$\omega_u = \frac{2\pi}{P_u}$$

and stop the experiment.

(1b) If the limit cycle is asymmetric,

   A. When the additional information, steady state gain and the magnitude of the load change, is needed, integrate system input and output responses ($\bar{u}$ and $\bar{e}$ in Equations 7.5 and 7.6) to find K_p and $K_L L$ according to Equations 7.33 and 7.34. Notice that, in order to have an accurate estimate of K_p and $K_L L$, one has to wait until the asymmetric response settles down. Typically, it takes two or three cycling periods.

   B. If the additional information is not needed, go to step 2.

(2) Perform an output-biased relay feedback test with the following bias value (Equation 7.39):

$$\delta_0^{(1)} = -\frac{h \Delta a^{(1)}}{a^{(1)}}$$

(2a) If the limit cycle becomes symmetric, then calculate K_u and ω_u

$$K_u = \frac{4h}{\pi a}$$

$$\omega_u = \frac{2\pi}{P_u}$$

and stop the experiment.

(2b) If the limit cycle is still asymmetric, then update the bias value in the output-biased relay according to

$$\delta_0^{(i+1)} = \delta_0^{(i)} - \frac{h \Delta a^{(i)}}{a^{(i)}} \qquad (7.40)$$

where $\delta_0^{(i+1)}$ stands for the bias value in the $(i+1)$th adjustment and $\Delta a^{(i)}$ represents the bias in the output with respect to the previous center point. It is recommended that the bias value is updated every one or tow oscillations.

7.4 Applications

Linear and nonlinear distillation examples are used to illustrate the effectiveness of the proposed system identification approach under load changes. Step-like and non-step types of load change are discussed. The ultimate gain and ultimate frequency are used to evaluate the correctness of the identified model.

7.4.1 Linear System

Consider the third-order plus dead time system:

$$G_3(s) = \frac{e^{-2s}}{(20s+1)(10s+1)(s+1)}$$

$$G_L(s) = \frac{e^{-s}}{(10s+1)}$$

Without load disturbance, a relay feedback experiment gives $K_u = 10.44$ and $\omega_u = 0.2126$ ($t = 0$–140 min in Figure 7.8). This corresponds to -5.6% and -2.9% errors in K_u and ω_u, respectively. When a step load change with $L = 0.8$ is introduced at $t = 140$ min, an asymmetric sustained oscillation results ($t = 140$–310 min in Figure 7.8). Again, K_u and ω_u can be estimated (Equations 7.5 and 7.6) and the corresponding errors are 7.46% and -31.54% respectively. Obviously, the quality of the estimates deteriorates as the result of load disturbance. At this point, one can proceed to step 1b-A to find additional information. From the system response we have $\Delta a = 0.08354$ and $a = 0.107$, while $\bar{u}$ and $\bar{e}$ can be found by integrating the input $u(t)$ and output $y(t)$. That gives $\bar{u} = -0.0942$ and $\bar{e} = -0.0942/0.1335 = -0.705$. Subsequently, the steady state gain and static load effect can be computed from Equations 7.33 and 7.34. The results are:

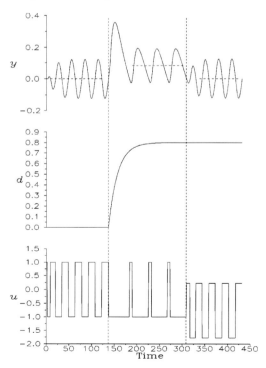

Figure 7.8. Input and output responses for the proposed output-biased relay feedback with and without load changes for G_3

$$K_P = \frac{-\bar{e}}{\left(\dfrac{\Delta a}{a}\right) \cdot h + \bar{u}}$$

$$= \frac{-(-0.705)}{\left(\dfrac{0.0835}{0.107}\right) - 0.0942}$$

$$= 1.03$$

Similarly, we find $K_L L = 0.803$.

If the information about K_p and $K_L L$ is not required, one can go directly to step 2. With the known values of Δa and a, the bias value δ_0 can be computed from Equation 7.39 and the result becomes $\delta_0 = 0.78$. Next, an output-biased relay feedback test is performed ($t > 300$ min in Figure 7.8) and K_u and ω_u can be found. This gives $K_u = 10.44$ and $\omega_u = 0.2126$, which are exactly the same as that of disturbance-free case. Therefore, the example clearly shows that, incorporated with Equation 7.39, the output-biased relay is very effective in maintaining the quality of the model in the face of load changes.

Despite the fact that chemical processes often face a progressive type of load disturbance, sometimes, non-step-like load changes occur. Again, the proposed method is tested against a series of step changes. For the third-order example, Figure 7.9 (middle line) shows the load changes. Once the asymmetry in the output is detected (Figure 7.9), the bias value δ_0 is adjusted at the end of each cycling period according to Equation 7.40. Despite the fact that the oscillation is hardly settling down (Figure 7.9), the estimates of K_u and ω_u can still be very accurate. Table 7.1 shows the estimated K_u and ω_u every three or four periods. The results clearly indicate that the proposed method can identify a quality model under a persistent load disturbance.

Table 7.1. Estimated K_u and ω_u under a series of step load changes at different oscillation periods

Period of oscillation	K_u	ω_u
3	10.20	0.2080
7	10.23	0.2102
10	10.25	0.2107
14	10.46	0.2125
17	10.25	0.2106
22	10.21	0.2105

152   Autotuning of PID Controllers

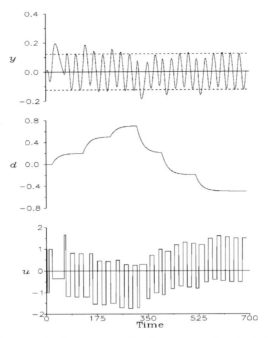

Figure 7.9. Input and output responses for the proposed output-biased relay feedback under a series of step load changes for G_3

7.4.2 Binary Distillation Column

Since the asymmetry in the output response can come from a load change as well as from process nonlinearity, the proposed approach is tested on a nonlinear process under load changes. A binary distillation column [8] is used to illustrate the effectiveness of the proposed approach. This is a 20-tray distillation column studied in Chapter 5. The product specifications are 98% and 2% of the light component on the top and bottoms respectively. The relative volatility is 2.26 with a reflux ratio of 1.76. Table 5.1 gives the steady state values. The control objective is to maintain the top and bottoms product compositions x_D and x_B by changing the reflux flow rate R and vapor boil-up rate V. This is the conventional $R-V$ control structure (Figure 5.12). In the nonlinear simulation, the following assumptions are made: (1) equal molar overflow, (2) 100% tray efficiency, (3) saturated liquid feed, (4) total condenser and partial reboiler and (5) perfect level control. An analyzer dead time of 6 min is used for top and bottoms composition measurements.

For this moderate-purity column, the stepping technique [9] is used to find the ultimate gain and ultimate frequency at the nominal operating point. This gives $K_u = 802.7$ and $\omega_u = 0.138$ for the $x_D - R$ loop. First, an ideal relay feedback test is performed on the $x_D - R$ loop and the results are $K_u = 736$ and $\omega_u = 0.132$ ($t = 0\text{--}210$ min in Figure 7.10). At $t = 210$ min, a -5% step change

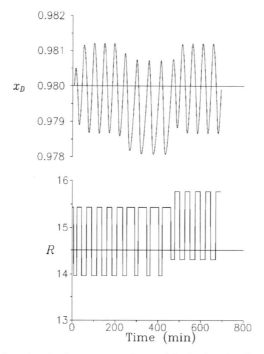

Figure 7.10. Input and output responses for an ideal relay feedback and an output-biased relay feedback under a +5% step feed flow rate change for the distillation example

in feed flow rate is introduced, and this leads to an asymmetric oscillation (Figure 7.10). The estimates of K_u and ω_u become 705.5 and 0.112 respectively. As a result of a feed flow rate disturbance, the estimation error in K_u increases from 8% to 12% and the error in the estimate of ω_u increases from 4% to 19%. From the system response (Figure 7.10), the parameters a and Δa can be read off immediately and the bias value of the output-biased relay can be calculated accordingly ($\delta_0 = 0.3328$). Next, the relay feedback test continues with an output-biased relay. The results show that ($t > 450$ min in Figure 7.10) symmetric output response is restored and the estimates of K_u and ω_u become 734.5 and 0.131 respectively. The results indicate that, under load disturbance, an improved estimate K_u and ω_u can be achieved using the output-biased relay. More importantly, the remedial action can be made by simply observing input–output response.

7.5 Conclusion

Any realistic autotuner should possess some ability to handle load disturbances. This is especially important for slow industrial processes. The reason is fairly obvious: load disturbance can give erroneous results in system identification. Despite the fact that the relay feedback test is more resilient to load changes, the estimates of K_u and ω_u deteriorate exponentially as the magnitude of the load increases. Since external load disturbance generally leads to an asymmetric oscillation, an intuitive approach is to restore a symmetric output response using the biased relay. In this chapter, a simple relationship is derived to find the bias value for the output-biased relay. The results show that good estimates of K_u and ω_u can be achieved under step-like and non-step-like load changes.

7.6 References

1. Smith CL. Digital computer process control. Intex: Scranton; 1972.
2. Seborg DE, Edgar TF, Mellichamp DA. Process dynamics and control. 2nd ed. New York: Wiley; 2004.
3. West JC, Douce JL, Livesley RK. The dual-input describing function and its use in the analysis of non-linear feedback systems. Proc. IEE. 1956;103B:463.
4. Oldenburger R, Boyer RC. Effects of extra sinusoidal input to nonlinear systems. Trans. ASME, 1962;84D:559.
5. Atherton DP. Nonlinear control engineering. London: Van Nostrand Reinhold; 1982.
6. Shen SH, Wu JS, Yu CC. Biased-relay feedback for system identification. AIChE J., 1996;42:1174.
7. Hang CC, Åström KJ, Ho WK. Relay auto-tuning in the presence of static load disturbance. Automatica, 1993;29:563.
8. Shen SH, Yu CC. Indirect feedforward control: Multivariable systems. Chem. Eng. Sci., 1992;47:3085.
9. Luyben WL. Process modeling, simulation and control for chemical engineers. 2nd ed. New York: McGraw-Hill; 1990.

8
Multiple Models for Process Nonlinearity

Intelligent control is now becoming common in the literature and in practice. Control systems with some types of intelligent features have begun to appear. Among these features, the abilities to perform automatic tuning in a multivariable environment and to adjust parameters as the operating condition changes are of primary importance in chemical process control. The reason is obvious: chemical processes, generally, are multivariable and nonlinear.

Chemical processes are often operated at different steady states. Changes in the operating condition are usually initiated by external factors. These parameters are often known *a priori*, *e.g.* changes in the production rate or product specification. The objective of process control is to achieve good transition while moving toward a new operating point and yet maintaining robust performance in the face of unknown disturbances. The concept of multiple models provides a useful framework for automated chemical process control [1–4]. Useful approaches can be found in a special issue edited by Johansen and Foss [5]. Since knowledge on process dynamics accumulates as the plant starts operation, provided with an efficient autotuning procedure, multiple models (or multiple sets of controller parameters) can be obtained in a straightforward manner. Conventionally, these models, if they exist, are utilized via a look-up table approach.

Here, we try to devise a framework for the control system design such that it works well over the entire operating regime. The automated control system design consists of two steps: (1) automatic tuning at a specific operating condition and (2) automatic model scheduling for the entire operating regime. The relay-feedback-based autotuning is proven reliable in the neighborhood of the nominal operating point. However, if the process is operated over a wide range of operating conditions, then the *local* controllers have to be retuned (as a result of large uncertainty bound) to meet a *global* performance criterion. Once multiple models are available, the next step is to employ the local model(s) at the corresponding operating condition. Approaches exist for incorporating models in different operating regimes. One is switching to a specific model if a certain condition is met (a crisp switching [4]). The other way is to combine local models using interpolation techniques (a fuzzy switching [6]). In this work, the fuzzy modeling of Takagi and Sugeno is used to schedule local models. It is a fuzzy augmentation of crisp models which provides a nice framework for model scheduling.

8.1 Autotuning and Local Model

Consider a relay-feedback system where $G(s)$ is the process transfer function, y is the controlled output, e is the error and u is the manipulated input. An ideal (on–off) relay is placed in the feedback loop. A relay of magnitude h is inserted in the feedback loop. Initially, the input u is increased by h. As the output y starts to increase (after a dead time D), the relay switches to the opposite position, $u = -h$. Because the phase lag is $-\pi$, a limit cycle with a period P_u results. The period of the limit cycle is the ultimate period. Therefore, the ultimate properties from this relay-feedback experiment are

$$\omega_u = 2\pi / P_u \tag{8.1}$$

$$K_u = 4h / \pi a \tag{8.2}$$

where h is the height of the relay and a is the amplitude of oscillation. Notice that the relay-feedback tests result in sustained oscillations for open-loop stable systems and most of the open-loop unstable systems.

Two approaches are taken here. One is the direct tuning using Ziegler–Nichols types of rule. Because of the multivariable nature of the process considered, we use the Shen–Yu tuning rule of Chapter 6. For a PI controller, we have $K_c = K_u / 3$ and $\tau_I = 2P_u$.

As will be shown later, in some cases transfer function models are preferable for the purpose of model scheduling. The ultimate gain K_u and ultimate frequency ω_u can be used directly to back-calculate the local transfer function model. As pointed out by several authors, the high-frequency characteristic of the integrator plus dead time model offers an attractive means in modeling *slow* chemical processes. The transfer functions have the following form:

$$G(s) = K_p e^{-Ds} / s \tag{8.3}$$

The model parameters can be solved directly from the ultimate gain and ultimate frequency:

$$K_p = \frac{\omega_u}{K_u} = \frac{2\pi}{K_u P_u} \tag{8.4}$$

$$D = \frac{\pi}{2\omega_u} = \frac{P_u}{4} \tag{8.5}$$

The controller parameters of the modified Ziegler–Nichols tuning can be expressed explicitly in terms of K_p and D. If the settings of Shen and Yu are used, then we have

$$K_c = \pi / 6K_p D \qquad (8.6)$$

$$\tau_I = 8D \qquad (8.7)$$

In this section, the relay-feedback test is introduced and steps required to perform the experiment are also given. Once you have obtained the information on the ultimate frequency, the controller settings can be decided using the modified Ziegler–Nichols methods. Moreover, the model parameters of the useful integrator plus dead time model can be found directly. This completes the tuning and modeling under a given operating condition. In other words, the local controller and local model can be found in a straightforward manner. It is very likely that, after some period of process operation, the autotuning procedure is repeated under different operating conditions. How can we utilize this local information to construct a global model?

8.2 Model Scheduling

Similar to gain scheduling, model scheduling is defined as using different process models as the operating condition changes. The output (or scheduled) variables z are often referred to as model parameters or controller settings and the input (or scheduling) variables x are the variables that indicate changes in the operating condition. They are often set by the operating condition, *e.g.* production rate, product specification, process outputs, *etc*. The model scheduling problem then becomes the following: given sets of process data (z,x), find the functions $z = f(x)$ which can describe the global behavior.

8.2.1 Takagi–Sugeno Fuzzy Model

The fuzzy modeling of Takagi and Sugeno [6] is employed to construct the global model. It uses fuzzy logic to interpolate between several models. A brief description of the fuzzy set is given. In the fuzzy set, a variable x may belong *partially* to a set (*e.g.* a set of high temperature). A membership function A characterizes this degree of belonging. A is defined as:

$$A(x): x \to [0,1], \quad x \in X$$

where X, generally, is a subset of R (real number) and the membership function falls between 0 and 1. The truth value (TV) of a proposition "x_1 is A_1 and x_2 is A_2" is expressed as

$$A_1(x_1) \wedge A_2(x_2) = \min(A_1(x_1), A_2(x_2))$$

where $\wedge$ is the logical AND operator.

Takagi and Sugeno [6] suggest that a multivariable system can be represented by the fuzzy implications $R^{(j)}$. Consider a multivariable system with n input variables (x_i, $i = 1, \cdots, n$) and one output z with k fuzzy implications.

$R^{(1)}$: If x_1 is $A_1^{(1)}, \cdots$ and x_n is $A_n^{(1)}$,

$$\text{then } z = p_0^1 + p_1^1 x_1 + \cdots + p_n^1 x_n$$

$R^{(k)}$: If x_1 is $A_1^{(k)}, \cdots$ and x_n is $A_n^{(k)}$,

$$\text{then } z = p_0^k + p_1^k x_1 + \cdots + p_n^k x_n$$

Then, the output z becomes

$$z = \sum_{j=1}^{k} \beta_j \left(p_0^j + p_1^j x_1 + \cdots + p_n^j x_n \right) \tag{8.8}$$

where

$$\beta_j = \frac{A_1^{(j)}(x_1) \wedge \cdots \wedge A_n^{(j)}(x_n)}{\sum_{j=1}^{k} \left[A_1^{(j)}(x_1) \wedge \cdots \wedge A_n^{(j)}(x_n) \right]} \tag{8.9}$$

In this work, the following assumptions are made: (1) the membership function is linear and (2) each regime (except two ends) is defined by two membership functions.

8.2.1.1 Single Input Systems

The Takagi–Sugeno method offers a general framework to establish a nonlinear (global) model between the scheduling variable x (*e.g.* production rate, product specification, *etc.*) and the scheduled variable z (*e.g.* process steady state gain, time constants, dead time, *etc.*). Let us use an SISO example to analyze the fuzzy model.

Example 8.1 SISO fuzzy model
Suppose the trend of the process variable z around two operating points is known. We have the following two implications:

$$R^{(1)}: \text{ If } x \text{ is } A^{(1)}, \text{ then } z = 0.1x + 0.9$$
$$R^{(2)}: \text{ If } x \text{ is } A^{(2)}, \text{ then } z = x + 1$$

The membership functions $A^{(1)}$ and $A^{(2)}$ are given in Figure 8.1, and the results show that the Takagi–Sugeno model leads to a piecewise nonlinear function between z and x. Analytically, the nonlinear function can be expressed as

$$z = r(x+1) + (1-r)(0.1x + 0.9), \quad 1 \le x \le 2 \tag{8.10}$$

where

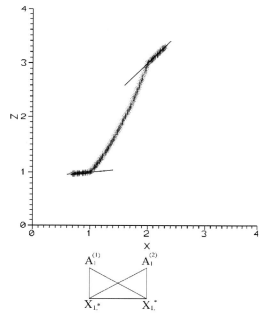

Figure 8.1. Membership functions and resultant global model from fuzzy modeling (Example 8.1)

$$r = \frac{x^* - x}{x^* - x_*} \qquad (8.11)$$

with x^* and x_* defining the upper and lower bounds of the regime. This is simply a linear combination of two linear functions, as shown in Figure 8.1. ∎

Several observations can be made immediately. Consider the linear membership functions in Figure 8.1 where the scheduling variable x superimposes the same range.

1. If the output variable z shows the same trend as the scheduling variable x varies (*i.e.* the slopes in the consequence of Figure 8.1 have the same sign), then the resultant nonlinear function is monotonic (*i.e.* the sign of the slope remains the same).

2. If the output variable z shows different trends as the scheduling variable x varies (*i.e.* the slopes have different signs), then the resultant nonlinear function is nonmonotonic.

An even simpler model scheduling mechanism can be devised. If we do not have any knowledge about the trend of the process variable (*i.e.* the slope in Figure 8.1), then the process variable can simply be set constant around the neighborhood where system identification is performed. Suppose the two data points we have are z^* at x^* and z_* at x_*. Mathematically, we have

$$z = rx^* + (1-r)x_*, \quad x_* \leq x \leq x^* \tag{8.12}$$

This is simply a linear interpolation between these two points. The following observation points out its limitation.

3. If the trend of the output variable z is not included, then the resultant function is simply a linear interpolation of these two different data points. It always exhibits monotonic behavior in between.

Actually, the general result is as follows: if the process description in the consequence is a polynomial with an order m, then the resultant function is also a polynomial function to the $m+1$ power. Despite its limitation, this simple approach offers an attractive alternative in most cases. Another nice feature of the Takagi–Sugeno modeling is that once a new identification result becomes available we can simply add another implication to the original rule sets. The function then becomes a piecewise linear function (*e.g.* Equation 8.12).

8.2.1.2 Multiple Inputs Systems

Systems with multiple scheduling variables are often encountered in practice. For example, both the production rate and the production specification are changed to meet the business condition (*e.g.* the Tennessee Eastman process [7] is a good example). Consider a general dual-input system with input variables x_1 and x_2 and one output variable z. Suppose we have four experimental results and the corresponding data are ($x_{1,*}$, $x_{2,*}$, $z^{(1,1)}$), (x_1^*, $x_{2,*}$, $z^{(2,1)}$), ($x_{1,*}$, x_2^*, $z^{(1,2)}$) and (x_1^*, x_2^*, $z^{(2,2)}$). Figure 8.2 gives the ranges of the two input variables and the membership functions. If these local data are employed in modeling, then again, the result of fuzzy implications can be expressed analytically. It becomes a *bilinear* function:

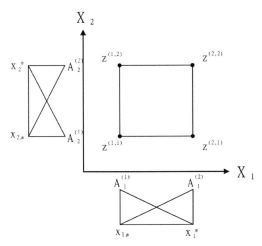

Figure 8.2. Linear membership functions for a two-input system

Multiple Models for Process Nonlinearity 161

$$z = r_1 r_2 z^{(1,1)} + r_1(1-r_2) z^{(1,2)} + (1-r_1) r_2 z^{(2,1)} + (1-r_1)(1-r_2) z^{(2,2)} \tag{8.13}$$

where

$$r_1 = \frac{x_1^* - x_1}{x_1^* - x_{1,*}} \quad \text{and} \quad r_2 = \frac{x_2^* - x_2}{x_2^* - x_{2,*}} \tag{8.14}$$

Example 8.2 MISO fuzzy model
Consider a system with two inputs x_1 and x_2 and one output z. Suppose the *trend* of the output variable is not known and plant tests give the following four data sets: $(x_1, x_2, z) = (1,1,1), (1,3,3), (1,3,3)$ and $(3,3,1)$. The four fuzzy implications are similar to that shown above with $x_{1,*} = x_{2,*} = 1$ and $x_1^* = x_2^* = 3$. Figure 8.3A shows the resultant bilinear function.

In practice, we may not have all the data points. For example, only three data points are available in Example 8.2, where $z^{(1,1)}$ corresponds to the nominal steady state, $z^{(2,1)}$ stands for an increase in the production rate and $z^{(1,2)}$ represents a change in the product specification. Under this circumstance, we only have three fuzzy implications ($R^{(1)}$, $R^{(2)}$ and $R^{(3)}$). The analytical expression then becomes

$$z = \frac{r_1 r_2}{r_1 + r_2 - r_1 r_2} z^{(1,1)} + \frac{r_1(1-r_2)}{r_1 + r_2 - r_1 r_2} z^{(1,2)} + \frac{(1-r_1) r_2}{r_1 + r_2 - r_1 r_2} z^{(2,1)} \tag{8.15}$$

With one less data point, the Takagi–Sugeno model gives a good description for the triangular region defined by $z^{(1,1)}$, $z^{(1,2)}$ and $z^{(2,1)}$. However, extrapolation outside this region is less reliable, as shown in Figure 8.3B. ∎

It is obvious that the extension of the Takagi–Sugeno model to a multivariable system is fairly straightforward. As expected, with the least process information, the

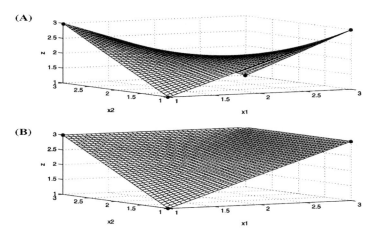

Figure 8.3. Global bilinear model from fuzzy modeling for Example 8.2 with (A) four data sets and (B) three data sets

162 Autotuning of PID Controllers

model leads to a bilinear system. However, one should be cautious when the model is extrapolated.

8.2.2 Selection of Scheduled Variable

From previous discussion, it becomes clear that the Takagi–Sugeno model interpolates *linearly* among data points. Hence, we need more than two data points to describe a function with nonmonotonic behavior. It generally requires more process information in quantity as well as in quality. Therefore, in building a global model, it is important to select appropriate scheduled variables z such that the nonmonotonic behavior can be avoided. Typical output variables in model scheduling are controller parameters and model parameters. It is rather intuitive to use the controller parameters (*e.g.* K_c and τ_I) as the output variables in the fuzzy modeling. Let us use the linear integrator plus dead time model to illustrate the effect of different scheduled variables. Suppose that T-L tuning is employed to tune the typical slow processes.

Consider the first case where both model parameters K_p and D increase as the operating condition changes (*i.e.* increase in the scheduling variable). Figure 8.4A shows that the controller parameters also change monotonically as the operating condition varies. However, a better global model can be achieved if the model parameters are selected as the scheduled variables. Numerically, it can be shown using the fuzzy modeling in the previous section for the case with or without a

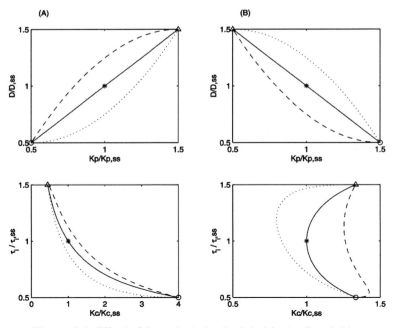

Figure 8.4. Effect of the selected scheduled (output) variables

process trend. The second case is that K_p and D change toward different directions as the operating condition changes (Figure 8.4B). This is a more likely situation in process systems. Since K_p represents the slope of the output responses and D is a measure of dead time, an increase in K_p and a decrease in D implies a faster output response. This is exactly the case in some plantwide control examples. However, if the controller parameters are used as the output variable, then we have a nonmonotonic behavior in the controller gain K_c, as shown in Figure 8.4B. As mentioned earlier, we need either more identification results or a very precise description of the process trend to find a reasonable global model. If the model parameters are employed as the scheduled variables, then only two data points are sufficient to construct a good global model. The examples clearly illustrate the importance in selecting the scheduled variables. For the integrator plus dead time model with the Ziegler–Nichols type of tuning, the model parameters seem to be a better choice, as the speed of response changes with the operating condition (this is most likely the case).

8.3 Nonlinear Control Applications

8.3.1 Transfer Function System

In this chapter, the integrator plus dead time model is chosen (Equation 8.3) to represent slow chemical processes. The controller settings of Equations 8.6 and 8.7 give a gain margin (GM) of 2.83 and phase margin (PM) of 46.1° for all possible model parameters (*i.e.* $K_p \neq 0$ and $D \neq 0$). First, we would like to know how well the nominal controller settings work. Considering the nominal condition of $\bar{K}_p = 1$ and $\bar{D} = 1$, Figure 8.5 shows the region of *robust stability* (RS). For example, the closed-loop system becomes unstable when $K_p = 2$ and $D = 2$ (Figure 8.5) and it remains stable for small values of K_p and D. Figure 8.5 shows that the settings remain stable for a fairly large region in the parameter space. A more useful assessment is that the region can achieve robust performance (RP). In this work, a very simple measure of RP is defined: a system is RP if and only if $2.21 \leq \text{GM} \leq 3.95$ and $36.1° \leq \text{PM} \leq 56.1°$. This means we allow $1/\text{GM}$ and PM to vary by ± 0.1 and $\pm 10°$ respectively. The following equations describing the magnitude M and phase ϕ are useful in finding the GM and PM as model parameters change. Substituting nominal tuning constants into the integrator plus dead time model, we have

$$M = \frac{\pi\sqrt{1+(8\omega\bar{D})^2}}{48(\omega\bar{D})^2} \frac{K_p}{\bar{K}_p} \quad (8.16)$$

$$\phi = -\pi - \frac{D}{\bar{D}}(\omega\bar{D}) + \tan^{-1}(8\omega\bar{D}) \quad (8.17)$$

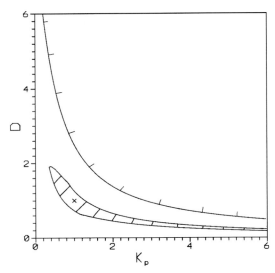

Figure 8.5. Regime of robust stability (RS) and robust performance (RP, hatched area)

where the overbar stands for the nominal condition. The region of the RP can then be found by solving Equations 8.16 and 8.17. The hatched area in Figure 8.5 indicates the parameter space where RP can be achieved. In other words, if the process drifts out of the hatched area the controller has to be retuned for good performance. Therefore, the region of RP can be used to evaluate the effectiveness of model scheduling approaches.

Suppose the process is operated at three different conditions: high, nominal and low productions, corresponding to $K_p = D = 0.5$, 1 and 2 respectively (indicated by × in Figure 8.6). We examined three approaches: (1) fixed gain control, (2) crisp switching control and (3) fuzzy switching control. By crisp switching we mean the model parameters (and consequently the controller parameters) are chosen from one of the three sets if a certain condition in the scheduling variable is met. Fuzzy switching implies the model parameters (and consequently the controller parameters) are generated from a fuzzy model (*e.g.* Equation 8.12). In the fixed gain control, we only have the nominal settings, the region of RP is indicated by the middle hatched area in Figure 8.6. Performance degradation can be expected as the operating point moves out of the region. If we choose to use the *crisp* model switching among three sets of model parameters, then, at best, the regions of RP are these three hatched areas. However, if the local models are scheduled according to the Takagi–Sugeno fuzzy implications, then we have a much larger region for RP, as shown in Figure 8.6. The degree of sophistication in fuzzy rules (*e.g.* with or without knowledge of process trend) has little effect on the RP region.

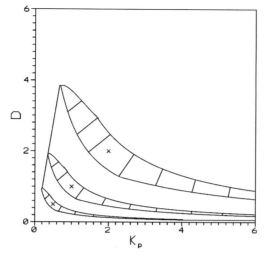

Figure 8.6. Regions of robust performance under different operating conditions (indicated by ×) for the fixed gain control (the middle hatched area), crisp switching (all three hatched areas) and fuzzy switching (the entire enclosed region)

Example 8.3 Nonlinear model
Consider the following nonlinear system:

$$y = \frac{K_p(y)e^{-D(y)s}}{s}u \qquad (8.18)$$

with

$$K_p(y) = y+1 \quad \text{and} \quad D(y) = y+1 \qquad (8.19)$$

Nominally, the system is operated at $y = 0$ and $u = 0$. A PI controller with Tyreus–Luyben tuning is employed and the results show that the fixed gain control gives oscillatory SP responses (dashed line in Figure 8.7). If we obtain new identification results at $y = 1$, then a fuzzy model scheduling can be constructed.

$$R^{(1)}: \quad \text{If } y \text{ is } A^{(1)}, \text{ then } K_p = 2 \text{ and } D = 2$$
$$R^{(2)}: \quad \text{If } y \text{ is } A^{(2)}, \text{ then } K_p = 1 \text{ and } D = 1$$

The membership functions are similar to that of Figure 8.1, except that the range of the scheduling variables y is between 0 and 1. The results show that much better SP responses can be obtained (solid line in Figure 8.7) when these two local models are scheduled using the simple Takagi–Sugeno fuzzy implications. Figure 8.8 shows the SP and load responses when the process is operated under different conditions. Here, a load transfer function of $1/(10s+1)$ and $L = 1$ are assumed.

∎

166 Autotuning of PID Controllers

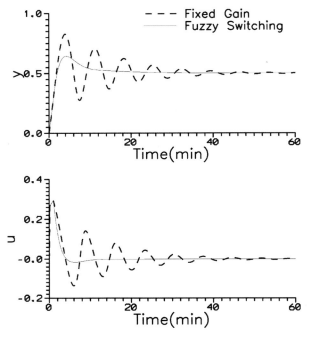

Figure 8.7. SP responses of Example 8.3 using the fixed gain control and fuzzy switching

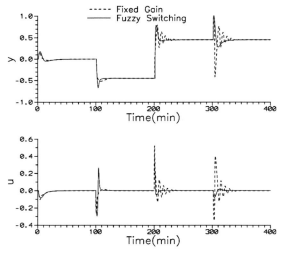

Figure 8.8. SP and load responses of Example 8.3 at different operating points using the fixed gain control and fuzzy switching

8.3.2 Tennessee Eastman Process

The Tennessee Eastman problem is a realistic complex reactor/separator process [7]. Several control strategies are proposed to solve the challenging problem. A detailed description of the process is given in Downs and Vogel [7]. The essential features of the process include an open-loop unstable reactor with two major reactions ($A+C+D \rightarrow G$ and $A+C+E \rightarrow H$), a separator removes unreacted light components and recycles them back to the reactor and a stripper further separates products from reactants (Figure 8.9). The temperature, pressure and levels are all interacting and nonlinear. The process is operated under different modes as the business condition changes. Mainly, we have to run the process under different production rates (PR) and different product specifications (PS, G/H mass ratios). Because we have a wide range of operating conditions, a single set of controller parameters is not expected to work well for the entire region. This is an ideal problem for the application of the multiple models.

The Luyben control structure [8] is employed (Figure 8.9). In this case, an on-demand product is set by the downstream process. The control structure consists of nine control loops: three level loops (reactor, separator and stripper), one pressure loop (reactor), three temperature loops (reactor, stripper and separator) and two composition loops. The integrating nature of the recycle structure leads to the use of simple proportional-only control on all loops. The reactor level and the stripper level are maintained by changing the inlet flow rates, the controllers require little tuning. The Luyben tuning constants are used for these two level loops and the pressure loops (Table 8.1). It should be noticed that the separator level is maintained by changing the cooling water flow; this manipulated input has strong effects on the separator temperature and pressure, as well as on the level. Therefore, care should be taken in the tuning of this level loop. Relay feedback tests are applied to the remaining six loops. The autotuning is carried out sequentially starting from the reactor temperature loop work to the two composition loops. The controller parameters are set to 1/3 of the ultimate gain, except for the stripper temperature loop (1/12). Table 8.1 gives the nominal settings.

Table 8.1. Nominal controller parameters for the Tennessee Eastman process

Loop	Unit	K_c	Transmitter span
Level	Reactor	4	100%
	Separator	2.35	100%
	Stripper	2	100%
Pressure	Reactor	3.33	3000 kPa
Temperature	Reactor	12.7	100°C
	Separator	0.96	100°C
	Stripper	108	100°C
Composition	A in recycle	115	100 mol%
	B in recycle	23.1	100 mol%

168 Autotuning of PID Controllers

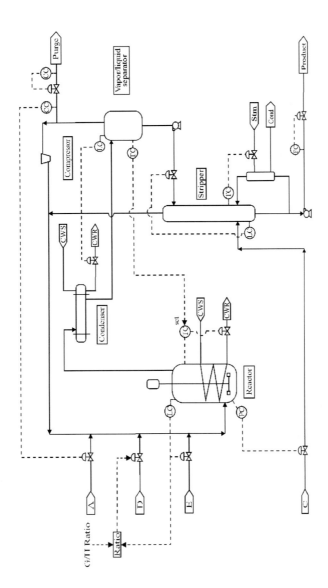

Figure 8.9. Tennessee Eastman process using Luyben control structure for the case of on-demand product

The model scheduling mechanism will become very complex if controller settings of all nine loops (or six loops) are scheduled. Therefore, it is important to devise a control structure such that only the minimum number of loops need to be retuned (actually, this can be viewed as a performance index of different control structures). Suppose, after some period of operation, we have performed relay feedback tests on ±30 production changes (PR = ±30%) with the nominal product specification (PS = 45/55). And we also have ultimate properties for 25/75 and 70/30 product ratios (PS = 25/75 and 70/30). Figure 8.10 shows that the ultimate gains stay fairly constant for most loops except for the separator level. Therefore, only the controller parameter for the separator level is scheduled. Because we have five data sets, the fuzzy implications thus become

$R^{(1)}$: If PR is PR* and PS is $\overline{PS}$, then $K = K^{(1,0)}$

$R^{(2)}$: If PR is $\overline{PR}$ and PS is $\overline{PS}$, then $K = K^{(0,0)}$

$R^{(3)}$: If PR is $\overline{PR}$ and PS is PS*, then $K = K^{(0,1)}$

$R^{(4)}$: If PR is $\overline{PR}$ and PS is PS$_*$, then $K = K^{(0,-1)}$

$R^{(5)}$: If PR is PR$_*$ and PS is $\overline{PS}$, then $K = K^{(-1,0)}$

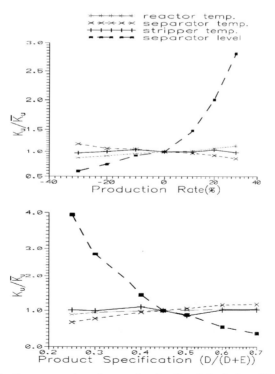

Figure 8.10. Controller parameters from relay feedback tests for different operating conditions: changes in the production rate and product specification

Figure 8.11 shows the corresponding membership functions. This is exactly the three data sets scenario described in Section 2, except that we have four triangular regions here. The results of the fuzzy modeling can also be expressed in the form of Equestion 8.15 (use three data sets for each region). Figure 8.12 shows the ultimate gain of the separator level loop as the production rate and product specification change. Provided with five data sets, the fuzzy implications allow us to move around different operating regions. For example, we have a simultaneous change in the production rate and the product specification ($PR = -20\%$ and $PS = 65/35$). Simulation results (Figure 8.13A and 8.13B) show that much better transient responses and better disturbance rejection (IDV(1) at time >10 h) are obtained using multiple local models. Moreover, this is achieved by scheduling only one level loop.

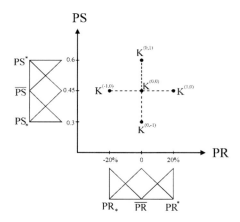

Figure 8.11. Linear membership functions for Tennessee Eastman process.

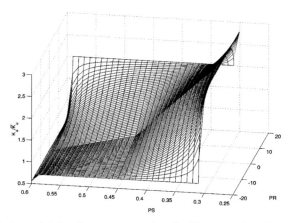

Figure 8.12. Global model for the separator level ultimate gain of the Tennessee Eastman process as production rate (PR) and product specification (PS) vary (−20% < PR< +20% and 0.3 < D/(D + E) < 0.6)

(A)

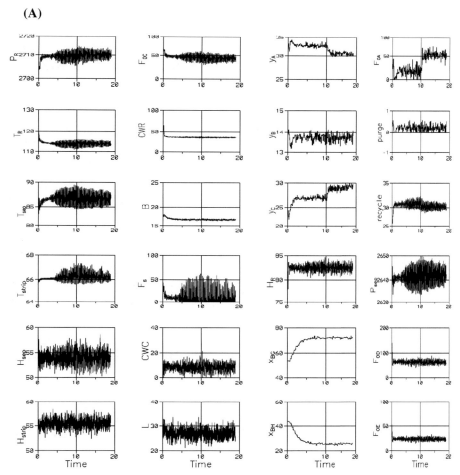

Figure 8.13A. Responses of the Tennessee Eastman process for simultaneous production rate and product specification changes followed by a load change (IDV(1) at time = 10 h) using: (A) fixed gain control

172 Autotuning of PID Controllers

(B)

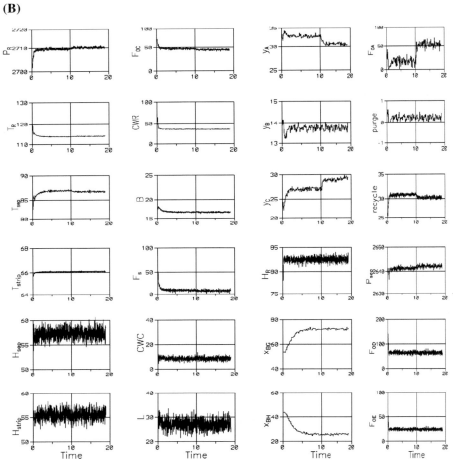

Figure 8.13B. Responses of the Tennessee Eastman process for simultaneous production rate and product specification changes followed by a load change (IDV(1) at time = 10 h) using: (B) fuzzy switching

8.4 Conclusion

In this chapter a framework for local autotuning and global model scheduling is proposed. Relay feedback is employed to find local models and then these models are scheduled using the Takagi–Sugeno fuzzy model. The characteristics of the resultant global model are analyzed. The importance of the selection of the scheduled parameters is emphasized. The proposed techniques are applied to simple transfer function models as well as large–scale recycle plants. Issues such as which variables should be selected and how many loops should be scheduled become important when dealing with large–scale systems. Simulation results show that improved performance can be achieved using relatively simple model scheduling.

8.5 References

1. Banerjee A, Arkun Y, Ogunnaike B, Pearson R. Estimation of nonlinear systems using linear multiple models. AIChE J. 1997;43:1204.

2. Chiu MS, Shan C. An IMC strategy using multiple models. Proc. 2nd KTS Symposium on Process System Engineering; Seoul; 1997.

3. Foss BA, Qin SJ. Interpolating optimizing process control. J. Process Control 1997;7:129.

4. Narendra KS, Balakrishnan J, Ciliz MK. Adaptation and learning using multiple models, switching and tuning. Control Syst. Mag. 1995;15:37.

5. Johansen TA, Foss BA. Multiple model approaches to modeling and control. Int. J. Control 1999;72:575.

6. Takagi T, Sugeno M. Fuzzy identification of systems and its applications to modeling and control. IEEE Trans. Syst. Man Cyber 1985;SMC-15:116.

7. Downs JJ, Vogel EF. A plantwide industrial process control problem. Comput. Chem. Eng. 1993;17:245.

8. Luyben WL. Simple regulatory control of the Eastman process. Ind. Eng. Chem. Res. 1996;35:3280.

9
Control Performance Monitoring

Controller performance assessment gives suitable information for developing improved controllers, should the performance be deemed unacceptable. Hitherto, much of the control research has been focused on developing model identification, controller design and for quantifying the robustness and performance of control systems at the design stage. Far less has been carried out on assessing the performance of the existing control systems. When the controller performance is inadequate, it is important to ascertain whether an acceptable level of performance can be achieved with the existing control structure. If this is possible, then one can take an appropriate course of action, like retuning of the existing controller, implementing an alternate control algorithm, *etc*. Where acceptable performance cannot be achieved with the existing control structure, then steps like implementing feedforward control, reducing the process dead time, adding new manipulated variables or sensors, *etc*. can be taken. Harris [1] and Desborough and Harris [2] have considered methods based on autocorrelation analysis that compare the existing control system performance with a minimum variance control standard. Åström [3] developed techniques for assessing the achievable performance using PID control in terms of bandwidth, normalized peak error for SP and load disturbances and rise time. This approach requires the Laplace transform model of the process. Swanda and Seborg [4] evaluated the performance of a controller for a step change in SP using settling time as the criterion. Huang and Jeng [5] assessed single-loop control systems with controllers of general/PI/PID structure, with IAE and rise time in tracking SP change. Huang and Shah [6] summarize the progress up to 1999.

Chiang and Yu [7] proposed a frequency-domain monitoring procedure based on the relay feedback for SISO systems. Ju and Chiu [8] further extended their work to multi-loop control systems. However, in both the above studies the relay experiments have to be carried out at least twice to evaluate the maximum closed-loop log modulus on-line. Here, we try to incorporate the shape factor (Chapter 4) into controller monitoring, such that a simpler procedure can be devised.

9.1 Shape-Factor for Monitoring

9.1.1 Shapes of the Relay Feedback

Consider a conventional feedback loop as shown in Figure 9.1. The process considered is a typical FOPDT process. The controller used is a PI controller. The open-loop process $G(s)$ and the controller $K(s)$ have the following general forms:

$$G(s) = \frac{K_p e^{-Ds}}{(\tau s + 1)} \tag{9.1}$$

$$K(s) = \frac{K_c (\tau_I s + 1)}{\tau_I s} \tag{9.2}$$

The product of $G(s)$ and $K(s)$ is known as the loop transfer function, designated as $Q(s)$ [9] and is given by

$$Q(s) = GK(s) = \frac{K_c K_p (\tau_I s + 1) e^{-Ds}}{\tau_I s (\tau s + 1)} = \frac{(\tau_I s + 1) e^{-Ds}}{\varepsilon s (\tau s + 1)} \tag{9.3}$$

where

$$\varepsilon = \frac{\tau_I}{K_c K_p} \tag{9.4}$$

An ideal relay is introduced in the feedback loop (before the controller) to conduct the relay feedback test. The shape of the relay feedback response depends on the value of the integral time τ_I of the PI controller. The mismatch in the integral time τ_I can also be observed from the shape of the response. Typically, we have three cases that arise based on the magnitude of τ_I when compared with that of the value of the time constant τ of the open-loop process.

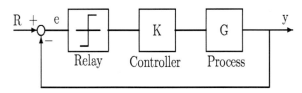

Figure 9.1. Simple feedback loop with relay

1. *Integral time is equal to the time constant* ($\tau_I/\tau = 1$).
   This is nothing but the original IMC-PI tuning rule [10]. With this integral time, the loop transfer function has a special structure (integrator plus dead time, IPDT) given by the following equation:

   $$Q(s) = GK(s) = \frac{K_c K_p e^{-Ds}}{\tau_I s} = \frac{e^{-Ds}}{\varepsilon s} \tag{9.5}$$

   The relay output response for the above process is shown in Figure 9.2. The step response of an integrator is a ramp function and that of a pure dead time process is just a step delay in time. Hence, the relay output response of the IPDT form of the loop transfer function is *triangular* in shape (a series of upward and downward ramps with a delay D). The half period of the relay response corresponds to dead time D, and the slope corresponds to $1/\varepsilon$ (see Chapter 4). Hence the peak corresponds to D/ε.

2. *Integral time is greater than the time constant* ($\tau_I/\tau > 1$).
   In this case the loop transfer function will be of the form presented in Equation 9.3 with a lead time constant τ_I greater than the lag time constant τ. The relay feedback response of such a transfer function demonstrates *convex* rise and fall, as shown in Figure 9.3.

3. *Integral time is less than the time constant* ($\tau_I/\tau < 1$).
   When the integral time takes a value less than the time constant, we have a transfer function in the form of Equation 9.3 with a lag time constant τ greater than the lead time constant τ_I. The relay feedback response displays a concave rise and fall shape. Depending on the sharpness of the shape, this case can be further classified into two different sub-cases. If the τ_I/τ ratio is greater than (τ_I/τ)critical then we can observe a sharp peak in the concave rise and fall, as shown in Figure 9.4a. On the other hand, if the τ_I/τ ratio is

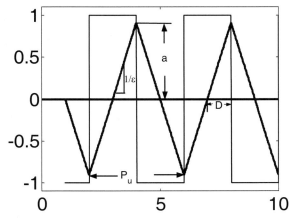

Figure 9.2. Relay output response of *GK(s)* for FOPDT process with $\tau_I = \tau$ (thick solid line) and the shifted input (thin solid line)

less than the $(\tau_I/\tau)_{\text{critical}}$ ratio, then we can observe a *rounded peak* in the concave rise and fall, as shown in Figure 9.4b. Generally, $(\tau_I/\tau)_{\text{critical}}$ is less than 0.5; the analytical expression in terms of model parameters will be derived later.

These observations are useful in the controller performance assessment using the shape factor from relay feedback.

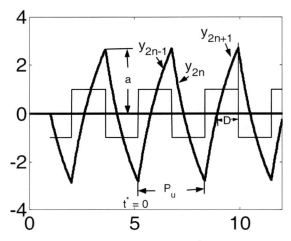

Figure 9.3. Relay output response of $GK(s)$ for FOPDT process with $\tau_I > \tau$ (thick solid line) and the shifted input (thin solid line)

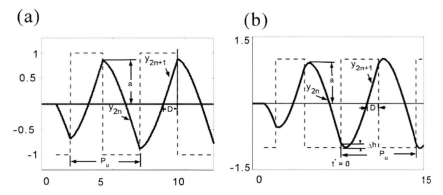

Figure 9.4. Relay output responses of $GK(s)$ for FOPDT process with $\tau_I < \tau$ (thick solid line) and the shifted input (thin solid line): (a) $(\tau_I/\tau) > (\tau_I/\tau)_{\text{critical}}$ and (b) $(\tau_I/\tau) < (\tau_I/\tau)_{\text{critical}}$

9.2 Performance Monitoring and Assessment

9.2.1 Optimal Performance

One of the finer ways of obtaining insight into the performance limitations is to consider an "ideal" controller that is resulting in optimal performance. By "optimal" we mean that the IAE is minimized for a unit SP change when a PI controller with $\tau_I = \tau$ is used. The performance measure J can be expressed as

$$J = \int_0^\infty |e(t)| \, dt \tag{9.6}$$

Here, ε/D (which can be viewed as the controller gain) is varied to locate the optimal performance. In Figure 9.5, J is normalized by J^*, where J^* is the minimum IAE corresponding to the achievable performance using inverse-based controllers and in this case it is simply the dead time (D) of the process, i.e. $J^* = D$. From the Figure 9.5, it is observed that, for a PI controller, $\varepsilon/D = 1.68$ results in the minimum J/J^* value of 2.1 and the shape of the relay output of $GK(s)$ for such a controller will be triangular, as shown in Figure 9.2.

Before getting into the monitoring details, we need to address the question: What is the difference between this monitoring method and the conventional direct autotuning? As pointed out by Luyben [11], from the shape of relay feedback responses, we have an idea about the model structure and corresponding tuning rule (*e.g.* ZN, TL, and IMC) can be applied directly. The proposed monitoring approach helps the operator to visualize whether the ultimate *performance* is achieved from the experiment. Let us take Figure 9.2 as an example. The ultimate PI performance for an FOPDT system corresponds to a triangular shape with a relay amplitude of $D/\varepsilon = D/(1.68D) = 1/1.68$ (with $h = 1$) (Figure 9.2). Any deviation from this shape indicates a potential problem in the controller setting. If the reset time is too large then we observe a convex rise in the relay response (Figure 9.3), and when the

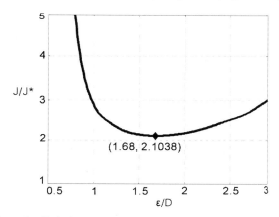

Figure 9.5. Effect of ε/D (τ_I/K_cK_p) on IAE for FOPDT processes with $\tau_I = \tau$ using PI controller

reset time is too small we find a concave rise in the relay response (Figure 9.4). The deviation of relay amplitude from 1/1.68 implies that the proportional gain has to be adjusted. This implies insights are gained from the relay feedback experiment, which is not possible with the conventional approach.

9.2.2 Proposed Monitoring and Assessment Procedure

Ongoing analyses reveal that the integral time τ_I affects the fundamental shape of the relay output responses, whereas the change in the controller gain K_c alters only the magnitude of the peak without affecting the shape. Therefore, the basic structure for the monitoring and assessment procedure involves: (1) identifying from the shape of the relay feedback the possible mismatch between the time constant and integral time, and employing the corresponding equations to find model parameters (e.g. τ, ε, and D); (2) adjusting the controller parameters according to the optimal PI settings (i.e. $\varepsilon/D = 1.68$ with $\tau_I = \tau$).

9.2.2.1 Case 1: $\tau_I/\tau = 1$

If the shape of the relay output test on $Q(s)$ is a triangle (as shown in Figure 9.2), then it is obvious that $Q(s)$ falls into this category. The procedure for assessing the controller performance of such a process is straightforward and involves the following steps (denoted as *Procedure 1* hereafter):

1. Read dead time D (time to the peak amplitude, e.g. Figure 9.2) and the peak amplitude a from the relay test.
2. Compute ε_{old} using the relation $\varepsilon_{old} = D/a$.
3. Find ε_{new} from the optimal performance chart, $\varepsilon_{new} = 1.68D$.
4. Adjust K_c according to $K_{c,new} = (K_{c,old})(\varepsilon_{old})(\varepsilon_{new})$ to acquire the optimal controller performance.

9.2.2.2 Case 2: $\tau_I/\tau > 1$

If the shape of the relay output test on $GK(s)$ is similar to that shown in Figure 9.3, then it is evident that $GK(s)$ falls into this category. Unlike the earlier case, the procedure for finding the value of ε_{old} is not straightforward. Analytical expressions are derived using time-domain analysis to represent the convex ascending and descending parts of Figure 9.3. Accordingly, the expressions for y_{2n} and y_{2n+1} can be given by Equations 9.7 and 9.8 respectively [12]:

$$y_{2n} = \left(\frac{P_u}{4\varepsilon}\right) - \left(\frac{t^*}{\varepsilon} + \left(\frac{\tau_I - \tau}{\varepsilon}\right) \times \left(1 - \frac{2e^{-t^*/\tau}}{1 + e^{(-P_u/2)/\tau}}\right)\right) \qquad (9.7)$$

Control Performance Monitoring   181

$$y_{2n+1} = -\left(\frac{P_u}{4\varepsilon}\right) + \left(\frac{t^*}{\varepsilon} + \left(\frac{\tau_I - \tau}{\varepsilon}\right) \times \left(1 - \frac{2e^{-t^*/\tau}}{1 + e^{(-P_u/2)/\tau}}\right)\right) \quad (9.8)$$

Parameters τ and ε can be estimated as follows. From Figure 9.3, it is clear that, when $t^* = 0$, $y_{2n+1} = -a$. Applying this boundary condition in Equation 9.8 and simplifying it we get

$$f_1(\tau,\varepsilon) = \frac{P_u}{4\varepsilon} - \left(\frac{\tau_I - \tau}{\varepsilon}\right)\left(1 - \frac{2}{1 + e^{(-P_u/2)/\tau}}\right) - a = 0 \quad (9.9)$$

Similarly, by substituting the boundary condition, $t^* = P_u/2 - D$ and $y_{2n+1} = 0$, in Equation 9.8 we get

$$f_2(\tau,\varepsilon) = \left(\frac{P_u/4 - D}{\varepsilon}\right) + \left(\frac{\tau_I - \tau}{\varepsilon}\right)\left(1 - \frac{2e^{-(P_u/2-D)/\tau}}{1 + e^{-(P_u/2)/\tau}}\right) = 0 \quad (9.10)$$

The two unknowns, namely τ and ε, can be determined by solving Equations 9.9 and 9.10. Thus, by knowing the values of τ and ε one can easily find the setting of the controller gain and reset time to obtain the optimal controller performance.

The following procedure, *Procedure 2*, describes the steps involved to obtain the optimal controller performance.

1. Read the dead time D (time to the peak amplitude, e.g. Figure 9.3), peak amplitude a and ultimate period P_u from the relay test.

2. Knowing τ_I from the controller settings and limit cycle data from step 1, the values of ε ($\varepsilon = \varepsilon_{\text{old}}$) and τ can be found using Equations 9.9 and 9.10.

3. Set $\tau_{I,\text{new}} = \tau$.

4. Find ε_{new} from the optimal performance chart, $\varepsilon_{\text{new}} = 1.68D$.

5. Adjust K_c according to $K_{c,\text{new}} = (K_{c,\text{old}})(\tau_{I,\text{new}} \varepsilon_{\text{old}})/(\tau_{I,\text{old}} \varepsilon_{\text{new}})$ to acquire the optimal controller performance.

The initial estimates for the two unknowns in Equations 9.9 and 9.10 are $\varepsilon = D/a$ and $\tau = 0.8\tau_I$.

9.2.2.3 Case 3: $\tau_I/\tau < 1$

In this case, after several numerical simulations, it was found that for every D/τ ratio there is a critical value for τ_I/τ ratio. Figure 9.6 gives the variation of the (τ_I/τ)critical ratio with the D/τ ratio for systems falling under this category. The implication of the (τ_I/τ)critical ratio can be explained as follows: consider a system with $GK(s)$ of the form given in Equation 9.3 having D, τ, ε, τ_I with values of 0.5, 1, 1 and 0.2 respectively. From Figure 9.6, the value of (τ_I/τ)critical for the system is found to be 0.2824. The relay feedback tests were conducted for the above system with the τ_I/τ ratio taking the values of 0.2, 0.2824 and 0.6

182 Autotuning of PID Controllers

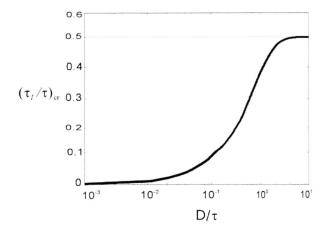

Figure 9.6. Variation of $(\tau_I/\tau)_{critical}$ ratio with D/τ ratio for FOPDT process with $\tau_I < \tau$

and the relay output responses thus obtained are displayed in Figure 9.7. From Figure 9.7 it is clear that for systems with a τ_I/τ ratio less than or equal to the $(\tau_I/\tau)_{critical}$ ratio the time elapsed from zero to peak value of $y(t)$, denoted as the apparent dead time D^*, becomes always greater than the true dead time D ($D^* > D$) and the shape of the relay output of $GK(s)$ will have a rounded peak (Figure 9.7(i) and (ii)). On the other hand, when the τ_I/τ ratio is greater than or equal to the $(\tau_I/\tau)_{critical}$ ratio the time elapsed from zero to peak value of $y(t)$ becomes always equal to D (i.e. $D^* = D$) and the shape of the relay output of $GK(s)$ will have a sharp peak (Figure 9.7 (iii)).

Case 3A: $(\tau_I/\tau)_{critical} \leq (\tau_I/\tau) < 1.$
As discussed earlier, if the relay output of $GK(s)$ is of the shape shown in Figure 9.4a or (Figure 9.7(iii)), then this implies that $\tau_I < \tau$ and the τ_I/τ ratio is greater than then $(\tau_I/\tau)_{critical}$ ratio. As pointed out earlier, the dead time of the process can be determined immediately, *i.e.* D is the time to the peak value; therefore, the same analytical expressions given in Equations 9.7 and 9.8 can be used to represent the concave descending and ascending parts of Figure 9.4a. In such a case, the assessment can be carried out by following the same procedure (Procedure 2) that is described for case 2, $(\tau_I/\tau) > 1$. That is, the two unknowns, namely τ and ε, can be determined by solving Equations 9.9 and 9.10. The values thus obtained can be used to find the settings of the controller gain to acquire optimal controller performance. Note that the initial estimate of τ is set to $1.2\tau_I$.

Case 3B; $(\tau_I/\tau) < (\tau_I/\tau)_{critical}.$
On the other hand, if the relay output of $GK(s)$ is of the shape shown in Figure 9.4b, then this implies that $\tau_I < \tau$ and the τ_I/τ ratio is less than the $(\tau_I/\tau)_{critical}$ ratio. In such a case, there is a minor modification in the assessment procedure. Similar to the earlier approach, Eqs 9.7 and 9.8 are used to represent the concave descending and ascending parts of Figure 9.4b. From Figure 9.4b we have

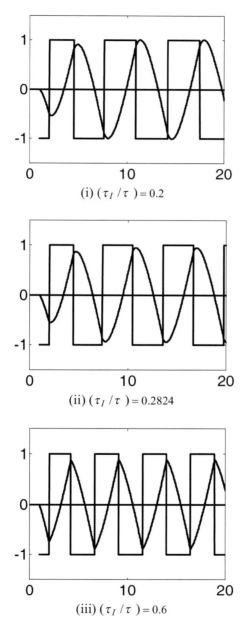

Figure 9.7. Relay feedback responses of *GK(s)* for FOPDT process with $\tau_I < \tau$ and $(\tau_I/\tau)_{critical} = 0.2824$ (solid line) and the relay input (dotted line); (i) $(\tau_I/\tau) < (\tau_I/\tau)_{critical}$, (ii) $(\tau_I/\tau) = (\tau_I/\tau)_{critical}$, (iii) $(\tau_I/\tau) > (\tau_I/\tau)_{critical}$

$$(y_{2n+1})_{t^*=P_u/2} - (y_{2n+1})_{t^*=0} + 2\Delta h = 2a \qquad (9.11)$$

After substituting various values in Equation 9.11 and simplifying it, at the boundary condition $t^* = \Delta D^*$ (time giving peak amplitude) and $y_{2n+1} = -a$ we have

$$f_1(\tau,\varepsilon) = \frac{P_u}{4\varepsilon} - \frac{\tau}{\varepsilon}\left(1 + \ln\left(\frac{2(\tau-\tau_I)}{\tau\left(1+e^{-(P_u/2)/\tau}\right)}\right)\right) + \left(\frac{\tau-\tau_I}{\varepsilon}\right) - a \qquad (9.12)$$

Similarly, by substituting the boundary condition $t^* = P_u/2 - D$ and $y_{2n+1} = 0$ in Equation 9.8 we get exactly the same expression as Equation 9.10. However, the dead time D is computed from the difference between the apparent dead time D^* and ΔD^*. That is:

$$D = D^* - \Delta D^* = D^* - \tau\left(\ln\left(\frac{2(\tau-\tau_I)}{\tau\left(1+e^{-(P_u/2)/\tau}\right)}\right)\right) \qquad (9.13)$$

Thus, the two unknowns, namely τ and ε, can be determined by solving Equations 9.10 and 9.12 when substituting Equation 9.13 for the dead time D. Equations 9.10 and 9.12 are non-linear and can be solved with the initial guess of ε from the slope of the corresponding relay feedback response ($\varepsilon = D/a$) and the initial guess of $\tau = 2\tau_I$.

The following, *Procedure 3*, are the various steps involved in assessing such a process:

1. Read the apparent dead time D^*, peak amplitude a and ultimate period P_u from the relay test.

2. Knowing τ_I from the controller settings, the values of ε_{old} and τ can be found using Equations 9.10 and 9.12.

3. Set $\tau_{I,\text{new}} = \tau$.

4. Obtain D from Equation 9.13.

5. Find ε_{new} from the optimal performance chart, $\varepsilon_{\text{new}} = 1.68D$.

6. Adjust K_c according to $K_{c,\text{new}} = (K_{c,\text{old}})(\tau_{I,\text{new}}\varepsilon_{\text{old}})/(\tau_{I,\text{old}}\varepsilon_{\text{new}})$ to acquire the optimal controller performance.

9.2.3 Illustrative Examples

The proposed scheme is illustrated for the FOPDT process for three cases, namely (τ_I/τ) equal to, greater than, and less than 1. For the example pertaining to $\tau_I < \tau$, the value of τ is taken as 1 and τ_I is taken as 50% of τ (Table 9.1). The (τ_I/τ)$_{\text{critical}}$ ratio for the system is 0.3873 (Figure 9.6). In other words, in the present example, the value of (τ_I/τ) ratio is greater than that of the (τ_I/τ)$_{\text{critical}}$ ratio. Hence, this example falls under Case 3A described in Section 9.2.2.3. An ideal relay feedback response for such a system and the closed-loop response with PI controller with original controller settings are shown in Figure 9.8a. The relay response shows a concave curve with sharp peaks. Hence, the procedure described in Section 9.2.2.3 for Case 3A (Procedure 2) is used to determine the new controller parameters. The estimated model, e.g. values of τ and ε_{old}, is very close to the true process (Table 9.1) and the PI controller is retuned to obtain the improved performance. The controller parameters and the value of $J/J_{\min}$ thus obtained are also presented in Table 9.1. The improvement in the controller performance is shown in Figure 9.8a. The value of $J/J_{\min}$ (=1) obtained for the above example is very much satisfactory and shows that the PI controller with the new controller settings exhibits almost optimal controller performance.

Another example is taken to illustrate Case 3B described in Section 9.2.2.3, where $\tau_I < \tau$ and the (τ_I/τ) ratio is less than the $(\tau_I/\tau)_{\text{critical}}$ ratio. The typical values of the $GK(s)$ for the example considered are given in Table 9.1. The value of τ is taken as 1 and τ_I is taken as 20% of τ. The $(\tau_I/\tau)_{\text{critical}}$ ratio for the system is 0.3873. In other words, the (τ_I/τ) ratio is less than the $(\tau_I/\tau)_{\text{critical}}$ ratio. Hence, this example falls under Case 3B of Section 9.2.2.3. Figure 9.8b shows the ideal relay feedback response and the closed-loop response with PI controller with original controller settings. The relay response shows a concave curve with round peaks. Hence, Procedure 3 described in Section 9.2.2.3 for Case 3B is used to determine the new controller parameters. The estimated values of τ, D, and K_p, the new controller parameters and the value of $J/J_{\min}$ thus obtained are also presented in Table 9.1. Figure 9.8b presents the closed-loop performance of the PI controller with the improved controller settings. In this case also the value of $J/J_{\min} = 1$ obtained reveals the accuracy of the proposed method.

The $GK(s)$ used to illustrate the case where $\tau_I/\tau = 1$ is also given in Table 9.1. The relay response and the closed-loop response with PI controller are shown in Figure 9.8c. The relay output is triangular in shape with sharp peaks. Hence, the procedure described in Section 9.2.2.1 (Procedure 1) is used to determine the new controller parameters (Table 9.1). Figure 9.8c also presents the closed-loop response of the PI controller with new controller settings. The controller offers a $J/J_{\min}$ value of 1, thus confirming the optimal controller performance.

For the example illustrating the case where $\tau_I/\tau > 1$, the various parameters of $GK(s)$ are given in Table 9.1. Figure 9.8d gives the relay output of $GK(s)$ for the example considered. The convex curve with a sharp peak suggests that the procedure described in Section 9.2.2.2 (Procedure 2) has to be used to determine the new controller parameters to obtain the optimal controller performance. Table 9.1 gives the values of the new controller parameters and the value of $J/J_{\min}$. The value of $J/J_{\min} = 1$ and the closed-loop performance with the new controller settings (Figure 9.8d) reveal that the PI controller with the new controller settings obtained using the above procedure yields the optimal controller performance.

Table 9.1. Estimation of PI controller tuning parameters of FOPDT process with $\tau_I < \tau$, $\tau_I = \tau$, and $\tau_I > \tau$

$G(s)$	$K(s)$	$G(s)^a$ (estimated)	Procedure^b	$\tau_{I,\text{NEW}}$	$K_{c,\text{NEW}}$	$J/J_{\min}$	Remarks
$1 > (\tau_I/\tau) > (\tau_I/\tau)_{\text{critical}}$							
$\dfrac{e^{-s}}{(s+1)}$	$0.5\left(\dfrac{0.5s+1}{0.5s}\right)$	$\dfrac{1.001e^{-s}}{(1.001s+1)}$	2	1.0009	0.5958	1	Concave Curved with sharp peaks (Figure 9.8a)
$1 > (\tau_I/\tau)_{\text{critical}} > (\tau_I/\tau)$							
$\dfrac{e^{-s}}{(s+1)}$	$0.2\left(\dfrac{0.2s+1}{0.2s}\right)$	$\dfrac{1.023e^{-0.995s}}{(1.000s+1)}$	3	1	0.5952	1	Concave Curved with rounded peaks (Figure 9.8b)
$(\tau_I/\tau) = 1$							
$\dfrac{e^{-s}}{(s+1)}$	$0.9\left(\dfrac{s+1}{s}\right)$	$\dfrac{0.999e^{-1.002s}}{(1.000s+1)}$	1	1	0.5952	1	Triangle with sharp peaks (Figure 9.8c)
$(\tau_I/\tau) > 1$							
$\dfrac{e^{-s}}{(s+1)}$	$4\left(\dfrac{4s+1}{4s}\right)$	$\dfrac{0.995e^{-1.001s}}{(0.992s+1)}$	2	0.9919	0.5904	1	Convex curve with sharp peaks (Figure 9.8d)

a The steady state gain back-calculated from ε.
b Computation procedure used.

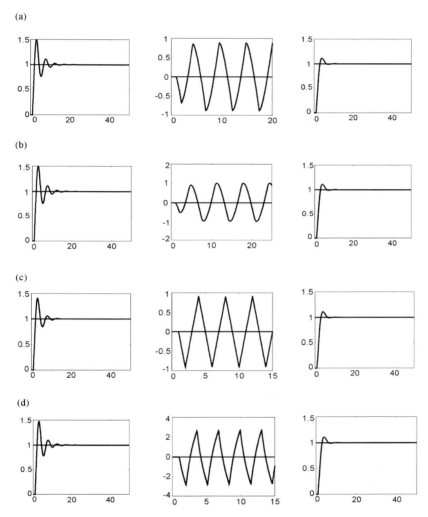

Figure 9.8. Performance assessment of FOPDT process for (a) $\tau_I < \tau$ and $(\tau_I/\tau) > (\tau_I/\tau)_{critical}$, (b) $\tau_I < \tau$ and $(\tau_I/\tau) < (\tau_I/\tau)_{critical}$, (c) $\tau_I = \tau$ and (d) $\tau_I > \tau$; left: closed-loop response with PI controller before adjustment; center: shape of the relay feedback response of $GK(s)$; right: closed-loop response with PI controller after adjustment

9.3 Applications

Up to now, we have examined the proposed method for processes with the same model structure (FOPDT); the same procedures will be tested against mismatches on model structures. Two model structures are employed here. One is the SOPDT system and the other is the HO system (order greater than 3).

9.3.1 Second-order Plus Dead Time Processes

The proposed scheme is extended for assessing the performance of a PI control scheme with an SOPDT process. The feedback loop shown in Figure 9.1 holds good in this case too, except that the open-loop process $G(s)$ has the following form:

$$G(s) = \frac{K_p e^{-Ds}}{(\tau s + 1)(\alpha \tau s + 1)} \quad (9.14)$$

The controller used is a PI controller with $K(s)$ as given in Equation 9.2. The product of $G(s)$ and $K(s)$ is given by Equation 9.15, where ε is same as that given in Equation 9.4

$$Q(s) = GK(s) = \frac{K_c K_p (\tau_I s + 1) e^{-Ds}}{\tau_I s (\tau s + 1)(\alpha \tau s + 1)} = \frac{(\tau_I s + 1) e^{-Ds}}{\varepsilon s (\tau s + 1)(\alpha \tau s + 1)} \quad (9.15)$$

Similar to the earlier case, relay feedback tests are conducted by introducing an ideal relay in the feedback loop before the controller and the shapes of the relay feedback responses are observed. Even in the case of an SOPDT process, we can find three cases, based on the magnitude of integral time τ_I compared with that of the value of $(1+\alpha)\tau$.

1. $\tau_I < (1+\alpha)\tau$. The relay feedback tests are conducted on $GK(s)$ for $\alpha = 0.2$, 0.6 and 1 with D/τ ratios of 0.5, 1, 5 and 10. The relay output responses thus obtained are displayed in Figure 9.9. The concave shape is observed in the first two rows. However, the third and fourth rows are of triangular shape with the sharpness of the peak increasing with increase in D/τ ratio.

2. $\tau_I = (1+\alpha)\tau$. The relay tests are conducted on $GK(s)$; the shapes of the relay output responses thus obtained are similar to those of Figure 9.9. As expected, we could observe the triangular shapes in the last two rows. Here, also, the sharpness of the peak increases with increase in D/τ ratio. However, in the first two rows the shapes resemble triangles with curved peaks.

3. $\tau_I > (1+\alpha)\tau$. Figure 9.10 displays the shapes of the relay feedback responses obtained for the SOPDT processes with different values of α and D/τ ratios. The convex shape is observed in the last two rows. However, in the first and second rows the shapes are of triangles with predominantly curved peaks.

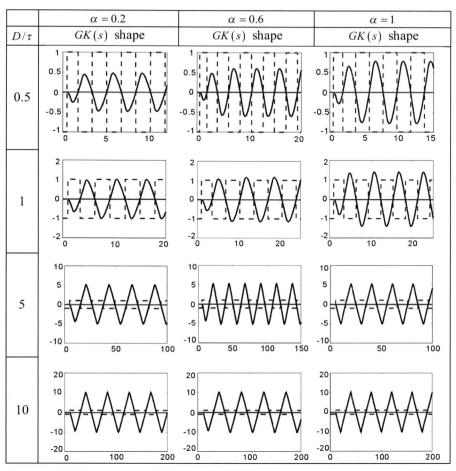

Figure 9.9. Relay feedback responses of GK(s) for SOPDT process with $\tau_I <$ $(1 + \alpha)\tau$ and the relay input

190 Autotuning of PID Controllers

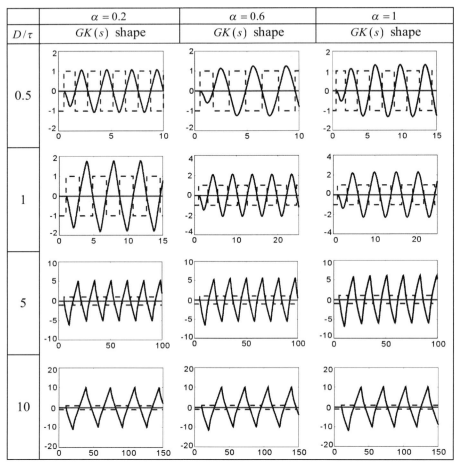

Figure 9.10. Relay feedback responses of $GK(s)$ for SOPDT process with $\tau_I > (1 + \alpha)\tau$ and the relay input

First, let us examine the case of $\tau_I = (1+\alpha)\tau$. In this case, three examples are considered, namely triangular shape with sharp peak (Figure 9.11a), triangular shape with partially curved peak (Figure 9.11b) and triangular shape with predominantly curved peak (Figure 9.11c) to illustrate the proposed scheme. The assessment is carried out using Procedure 1 given in Section 9.2.2.1, except that the value of ε_{new} is taken as $1.7D$ instead of $1.68D$ to have uniformity for the three cases. Even though the closed-loop responses obtained after necessary correction (Figure 9.11a–c) are satisfactory, better responses could be achieved in Figure 9.11b and c by taking slightly higher values of ε_{new} ($1.9D$ and $2D$ respectively for Figure 9.11b and c). The values of $G(s)$ and $K(s)$ used for illustration, the estimated controller parameters and also the values of J/J_{min} given in Table 9.2 are satisfactory.

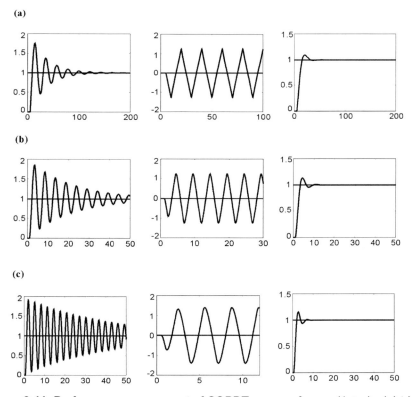

Figure 9.11. Performance assessment of SOPDT process for $\tau_I = (1 + \alpha)\tau$. (a) triangular shape with sharp peak, (b) triangular shape with partially curved peak, and (c) triangular shape with predominantly curved peak. Left: closed-loop response with PI controller before adjustment; center: shape of the relay feedback response of $GK(s)$; right: closed-loop response with PI controller after adjustment

Table 9.2. Estimation of PI controller tuning parameters of SOPDT process with $\tau_I = (1+\alpha)\tau$, $\tau_I < (1+\alpha)\tau$ and $\tau_I > (1+\alpha)\tau$

$G(s)$	$K(s)$	$G(s)$ [a]	Procedure [b]	$\tau_{I,\text{NEW}}$	$K_{c,\text{NEW}}$	$J/J_{\min}$ [c]	Remarks
$\tau_I = (1+\alpha)\tau$							
$\dfrac{e^{-5s}}{(s+1)(0.2s+1)}$	$0.3\left(\dfrac{1.2s+1}{1.2s}\right)$	$\dfrac{0.988\,e^{-5.12s}}{(1.2s+1)}$	1	1.2	0.1379	1.0013	Sharp peaks (Figure 9.11a)
$\dfrac{e^{-1s}}{(s+1)(0.4s+1)}$	$1.4\left(\dfrac{1.4s+1}{1.4s}\right)$	$\dfrac{1.021\,e^{-1.22s}}{(1.4s+1)}$	1	1.4	0.6750	1.2201	Partially curved peaks (Figure 9.11b)
$\dfrac{e^{-0.5s}}{(s+1)(0.4s+1)}$	$1.4\left(\dfrac{1.4s+1}{1.4s}\right)$	$\dfrac{0.946\,e^{-0.74s}}{(1.4s+1)}$	1	1.4	1.1128	1.0161	Predominantly curved peaks (Figure 9.11c)
$\tau_I < (1+\alpha)\tau$							
$\dfrac{e^{-1s}}{(s+1)(0.2s+1)}$	$0.6\left(\dfrac{0.6s+1}{0.6s}\right)$	$\dfrac{0.940\,e^{-1.27s}}{(1.0269s+1)}$	2	1.0269	0.4756	1.0806	Concave Curve (Figure 9.12a)
$\tau_I > (1+\alpha)\tau$							
$\dfrac{e^{-5s}}{(s+1)(0.2s+1)}$	$0.6\left(\dfrac{2.4s+1}{2.4s}\right)$	$\dfrac{0.895\,e^{-5.06s}}{(0.596s+1)}$	3	0.5963	0.0693	1.1212	Convex curve (Figure 9.12b)

a The steady-state gain back-calculated from ε.
b Computation procedure used.
c Compared to minimum IAE ($J_{\min}$) using PI controller with $\tau_I = (1+\alpha)\tau$.

Consider the case of $\tau_I < (1+\alpha)\tau$. The example used to illustrate this case is also given in Table 9.2. The shape of the relay feedback response of $GK(s)$ is shown in Figure 9.12a. Since the shape resembles a concave curve, the procedure given in Section 9.2.2.3 (case A) is used (with ε_{new} as $1.7D$ instead of $1.68D$) to perform the assessment of the controller. The estimated controller parameters and the value of J/J_{min} are given in Table 9.2. The closed-loop response obtained after adjustment (Figure 9.12a) and the value of J/J_{min} (Table 9.2) reveal the optimal controller performance of the controller.

For the case of $\tau_I < (1+\alpha)\tau$, when the shape of the relay response of $GK(s)$ is a convex curve (Figure 9.12b), the procedure given in Section 9.2.2.2 can be used (with ε_{new} as $1.7D$ instead of $1.68D$) to perform the controller assessment. The details of $G(s)$ and $K(s)$ used for illustration, the estimated controller parameters and also the values of J/J_{min} are tabulated in Table 9.2. The closed-loop performance of the controller after the adjustment is displayed in Figure 9.12b.

Thus, Figures 9.11 and 9.12 and the tabulated results of Table 9.2 justify the extension of the proposed method for SOPDT processes.

9.3.2 High-order Processes

The proposed performance monitoring and assessment scheme is extended for assessing the performance of the PI control scheme with HO processes. The arrangement of the feedback loop is similar to that shown in Figure 9.1. However,

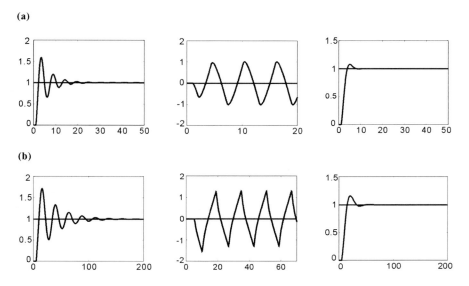

Figure 9.12. Performance assessment of SOPDT process for (a) $\tau_I < (1 + \alpha)\tau$ and (b) $\tau_I > (1 + \alpha)\tau$; left: closed-loop response with PI controller before adjustment; center: shape of the relay feedback response of $GK(s)$; right: closed-loop response with PI controller after adjustment

the open-loop process $G(s)$ has the following form:

$$G(s) = \frac{K_p e^{-Ds}}{(\tau s+1)^n} \tag{9.16}$$

with $n \geq 3$. The controller used is a PI controller with $K(s)$ as given by Equation 9.2. The product of $G(s)$ and $K(s)$ is given by Equation 9.17, where ε is same as that given in Equation 9.4:

$$GK(s) = \frac{K_c K_p (\tau_I s+1) e^{-Ds}}{(\tau_I s)(\tau s+1)^n} = \frac{(\tau_I s+1) e^{-Ds}}{(\varepsilon s)(\tau s+1)^n} \tag{9.17}$$

Three different examples are taken to illustrate the assessment procedure for the three cases of the HO process.

First consider an example of $\tau_I < n\tau$. Figure 9.13a displays the shape of the relay feedback response when $\tau_I < n\tau$. Though a concave shape is expected, we could observe only a shape almost resembling a symmetrical triangle with predominantly curved peaks. Hence, the procedure presented in Section 9.2.2.1, Procedure 1, is used (instead of Procedure 3 given in Section 9.2.2.3) with ε_{new} as $1.7D$. The closed-loop response obtained after controller parameter adjustment is also shown in Figure 9.13a. The values of $G(s)$ and $K(s)$ used for illustration, the estimated controller parameters and the value of $J/J_{\min}$ are tabulated in Table 9.3.

The relay feedback response of $GK(s)$ of the HO process when $I = n\tau$ is shown in Figure 9.13b. As expected, the shape is triangular, and hence Procedure 1 presented in Section 9.2.2.1 is used with ε_{new} as $1.7D$. Table 9.3 gives the details of $G(s)$, $K(s)$, the estimated controller parameters and the value of $J/J_{\min}$. Figure 9.13b displays the closed-loop response of the PI controller after adjustment.

In this case, the $GK(s)$ is taken such that $\tau_I > n\tau$ and the relay feedback response obtained from the relay test is shown in Figure 9.13c. The shape is a convex curve and matches with our intuition. Hence, Procedure 2 described in Section 9.2.2.2 is used to assess the performance of the controller. Figure 9.13c also displays the closed-loop response of the PI controller after adjustment. The details of $G(s)$ and $K(s)$ used in the illustration, the estimated controller parameters and the value of $J/J_{\min}$ are tabulated in Table 9.3.

Figures 9.13a–c and the values of $J/J_{\min}$ of Table 9.3 reveal that the proposed method can also be extended for HO processes. In fact, the proposed method works well for a third-order process, but as the order increases (fifth order and above) the proposed method offers satisfactory values of $J/J_{\min}$ for small D/τ ratio only.

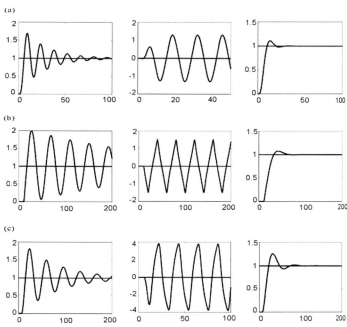

Figure 9.13. Performance assessment of higher order processes: (a) fifth-order process for $\tau_I < n\tau$; (b) third-order process for $\tau_I = n\tau$; (c) fifth-order process for $\tau_I > n\tau$; left: closed-loop response with PI controller before adjustment; center: shape of the relay feedback response of $GK(s)$; right: closed-loop response with PI controller after adjustment

Table 9.3. Estimation of PI controller tuning parameters of HO process with $\tau_I = n\tau$, $\tau_I > n\tau$ and $\tau_I < n\tau$

$G(s)$	$K(s)$	$G(s)^a$ estimated	Procedure[b]	$\tau_{I,\text{NEW}}$	$K_{c,\text{NEW}}$	$J/J_{\min}^{\ c}$	Remarks
$\tau_I < n\tau$							
$\dfrac{e^{-0.5s}}{(s+1)^5}$	$1.25\left(\dfrac{2.5s+1}{2.5s}\right)$	$\dfrac{0.802\,e^{-3.26s}}{(2.5s+1)}$	1	2.5	0.4511	1.0003	Curved peaks (Figure 9.13a)
$\tau_I = n\tau$							
$\dfrac{e^{-10s}}{(s+1)^3}$	$0.43\left(\dfrac{3s+1}{3s}\right)$	$\dfrac{0.969\,e^{-10.84s}}{(3s+1)}$	1	3	0.1628	1.0259	Sharp peaks (Figure 9.13b)
$\tau_I > n\tau$							
$\dfrac{e^{-5s}}{(s+1)^5}$	$2\left(\dfrac{10s+1}{10s}\right)$	$\dfrac{0.813\,e^{-6.57s}}{(1.420s+1)}$	3	1.4198	0.1271	1.2688	Convex curve (Figure 9.13c)

a The steady state gain back-calculated from ε.
b Computation procedure used.
c Compared to minimum IAE ($J_{\min}$) using PI controller with $\tau_I = n\tau$.

9.4 Conclusion

A method for performance monitoring and assessment of a single-loop system, using the shape factor from relay feedback, is presented. It was shown that the shape of the relay output characterizes the performance of the controller. Improved controller settings can be back-calculated whenever the shape of the relay feedback deviates from its optimal form. The results show that the proposed scheme provides a reliable way to assess controller performance, and, if necessary, to readjust the controller parameters. More importantly, it employs only one relay test and provides a simple method, compatible even with a non-expert operator, to assess the controller performance as well as to find the correct controller parameters.

9.5 References

1. Harris TJ. Assessment of control loop performance. Can. J. Chem. Eng. 1989;67:856.

2. Desborough L, TJ. Harris. Performance assessment measures for univariate feedback control. Can. J. Chem. Eng. 1992;70:1186.

3. Åström KJ. Assessment of achievable performance of simple feedback loops. Int. J. Adapt Control Signal Process 1991;5:3.

4. Swanda AP, Seborg DE. Controller performance assessment based on SP response data. Proc. of A.C.C.; San Diego, California; 1999.

5. Huang HP, Jeng JC. Monitoring and assessment of control performance for single loop systems. Ind. Eng. Chem. Res. 2002;41:1297.

6. Huang B, Shah SL. Performance assessment of control loops: Theory and applications. London: Springer-Verlag; 1999.

7. Chiang RC, Yu CC. Monitoring procedure for intelligent control: On-line identification of maximum closed loop log modulus. Ind. Eng. Chem. Res. 1993;32:90.

8. Ju J, Chiu MS. Relay based on line monitoring procedures for 2×2 and 3×3 multiloop control systems. Ind. Eng. Chem. Res. 1997;36:2225.

9. Skogestad S, Postlethwaite I. Multivariable feedback control. New York: Wiley; 1996.

10. Rivera DE, Morari M, Skogestad S. Internal model control. 4. PID controller design. Ind. Eng. Chem. Process Des. Dev. 1986;25:252.

11. Luyben WL. Getting more information from relay feedback tests. Ind. Eng. Chem. Res. 2001;40:4391.

12. Thyagarajan T, Yu CC, Huang HP. Assessment of controller performance: A relay feedback approach. Chem. Eng. Sci. 2003;58:497.

10

Imperfect Actuators

As mentioned in Chapter 1, an imperfect actuator (*e.g.* control valve) is a major factor in poor control performance [1].

Recent years have seen renewed interest in the study of linear systems in the presence of imperfect actuators. Adaptive schemes and fuzzy control are proposed for dead-zone and/or hysteresis compensation [2–4]. Mechanical motion control is a typical area of application. Less attention is paid to pneumatic actuators [5,6]. The problem of imperfect actuators can become very severe when autotuners are employed to find controller parameters. The standard autotuning procedure can lead to erroneous results and the performance of the control loops degrades drastically [7]. For example, an ideal relay feedback test tends to overestimate the ultimate gain for valves with hysteresis and this, subsequently, leads to oscillatory or unstable closed-loop responses [6,8].

This chapter studies quantitatively the estimation errors (in K_u and ω_u) for valves with hysteresis under relay feedback. Comparisons are also made between ideal relay and the saturation (ramp-type) relay and an approach for simultaneous identification of ultimate properties and width of hysteresis is proposed.

10.1 Potential Problems

Recall that, for an ideal relay, the ultimate properties from the relay feedback experiment are

$$\omega_u = \frac{2\pi}{P_u} \qquad (10.1)$$

$$K_u = \frac{4h}{\pi a} \qquad (10.2)$$

where h is the height of the relay and a is the amplitude of oscillation. For the saturation relay (Figure 10.1B), K_u can be computed from

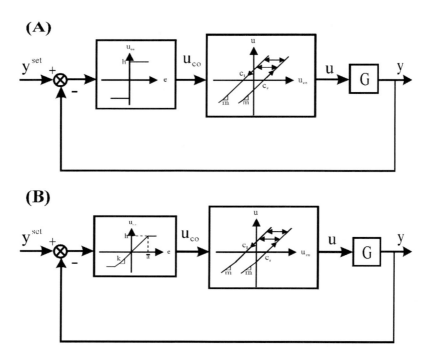

Figure 10.1. Actuator with hysteresis under (A) ideal relay feedback and (B) saturation relay feedback

$$K_u = \frac{2h}{\pi \bar{a}}\left[\left(\sin^{-1}\frac{\bar{a}}{a}\right)+\left(\frac{\bar{a}}{a}\sqrt{1-\left(\frac{\bar{a}}{a}\right)^2}\right)\right] \qquad (10.3)$$

where $\bar{a} = h/k$.

Consider an imperfect actuator in a feedback loop with the process $G(s)$ and a controller $K(s)$. An imperfect actuator is characterized by the hysteresis width and a slope as shown in Figure 10.1. The input and output of the hysteresis, u_{co} and u, are described by

$$u_{co}(t) = \begin{cases} m(u(t)-c_l), & u(t) < c_l + \dfrac{u_{co}(t-1)}{m} \\ m(u(t)-c_r), & u(t) < c_r + \dfrac{u_{co}(t-1)}{m} \\ u(t-1), & c_l + \dfrac{u_{co}(t-1)}{m} \leq u(t) \leq c_r + \dfrac{u_{co}(t-1)}{m} \end{cases} \qquad (10.4)$$

where t is the index of time, u_{co} is the controller output, u is the actual position of the actuator, m is the slope of hysteresis and c_r and c_1 (negative) are constants describing the width of the hysteresis. Here, we assume $m=1$ and $|c_1|=|c_r|$. The width of the hysteresis (dead band, DB) is defined as

$$DB = c_r - c_1 \tag{10.5}$$

Let us use a typical process transfer function to illustrate the effect of actuator hysteresis on ultimate properties. This is an *nth*-order plus dead time system.

$$G(s) = \frac{K_p e^{-Ds}}{(\tau s+1)^n} \tag{10.6}$$

First let $n=1$, a first-order system, and a time constant $\tau = 20$ and steady state gain $K_p = 1$ are assumed. Two types of relay are considered: an ideal relay and a saturation relay (Figure 10.1). A relay height h of 5% is used in both cases. For the saturation relay, the slope k is taken as $1.4 K_u$ to ensure a successful test (Figure 10.1B). Figure 10.2 shows the relative error in K_u and P_u as the width of hysteresis (DB) changes. As expected, the error in K_u increases as DB increases for both cases. When DB reaches 5%, the ideal relay feedback *overestimates* K_u by a factor of 2! That means a relay-feedback autotuner with Ziegler–Nichols tuning can produce an unstable closed-loop system if DB is greater than 5%. The saturation relay shows some improvement on the estimation error for K_u. For the ultimate period, the ideal relay produces a correct estimate and the saturation relay, on the other hand, shows a maximum error of 35.71% (Figure 10.2) for a DB of 7% (Figure 10.2). Notice that an overestimate in P_u tends to give a more sluggish response, since we have a larger reset time τ_I. Figure 10.3 shows how the estimation errors vary with the dead time D/τ. For the ideal relay, the estimation errors remain the same as D/τ changes. The saturation relay, however, shows a smaller estimation error in K_u for the system with a smaller D/τ ratio (*i.e.* systems with a long time constant). The results are just the opposite for the estimation of P_u, as shown in Figure 10.3. Furthermore, we can improve the estimation in K_u by reducing the slope of the saturation relay, as shown in Figure 10.4. However, it should be emphasized that if the slope $k < K_u$ is too small, then the relay feedback will fail to generate a sustained oscillation.

Up to this point the results come from the study of FOPDT systems. On some occasions, higher order systems are encountered and third-order systems are used to illustrate the effects of hysteresis. Figure 10.5 reveals that hysteresis leads to an overestimation in the ultimate gain and the offsets in the estimation are exactly the same as that of the first-order system (*i.e.* Figure 10.2) for the ideal relay. Qualitatively similar estimation errors are observed for the saturation relay, as shown in Figure 10.5 (*e.g.* comparing dashed lines in Figures 10.2 and 10.5). Actually, this is within one's expectation, since hysteresis results in a discount in the relay height and, without any compensation, this simply overestimates K_u. The results can be summarized as follows.

200 Autotuning of PID Controllers

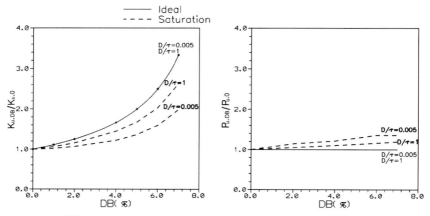

Figure 10.2. Effects of hysteresis width (DB) on estimation errors for ideal and saturation relays for first-order system with different D/τ ratios

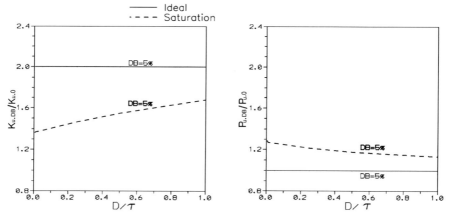

Figure 10.3. Effects of dead time to time constant (D/τ) ratios on estimation errors for ideal and saturation relays for first-order system with DB = 5%

(1) Relay feedback tests overestimate K_u for actuators with hysteresis and relative errors increase with the DB. This is generally true, especially for the ideal and saturation relays (Figures 10.2 and 10.5).

(2) The saturation relay improves the estimation error in K_u over the ideal relay. Moreover, the smaller the slope, the larger the margin of improvement (Figure 10.4).

(3) The saturation relay gives a better estimate in K_u for systems with a small D/τ ratio (Figure 10.3).

(4) For an ideal relay, the estimation error of K_u remains the same for different D/τ ratios and the estimate of P_u is not affected by the hysteresis (Figure 10.3).

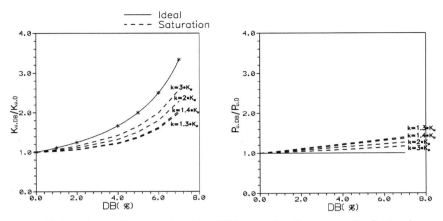

Figure 10.4. Effects of hysteresis width (DB) on estimation errors for first-order system with the saturation relays using different slopes k with $D/\tau = 0.005$

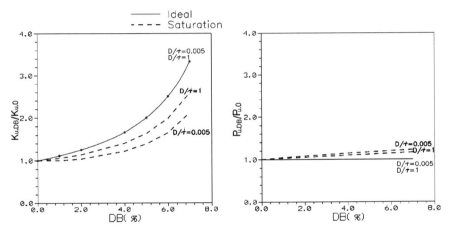

Figure 10.5. Effects of hysteresis width (DB) on estimation errors for ideal and saturation relays for third-order system with different D/τ ratios

Actually, for the ideal relay, the estimation error of K_u can be expressed analytically in terms DB and h. By comparing the ultimate gain using the assumed and true relay heights, the relative error in the estimation of K_u can be computed. From Equation 10.2 we have

$$\frac{K_{u,\text{DB}}}{K_u} = \frac{h}{h - \frac{\text{DB}}{2}} \qquad (10.7)$$

where K_u is the ultimate gain without hysteresis and $K_{u,\text{DB}}$ is the ultimate gain under a hysteresis actuator with a width DB. Since the true relay height is lowered by a factor of $\text{DB}/2$, we subsequently overestimate the steady state gain as well as the ultimate gain (the information is concealed by the hysteresis). Despite the fact that the saturation relay is more robust with respect to the hysteresis actuator, the estimation error of K_u can still be as large as 60% for a first-order system with $D/\tau = 1$ and $DB = 5\%$. Therefore, it is necessary to provide remedial action for systems with imperfect actuators.

10.2 Identification Procedure

10.2.1 Two-step Procedure

Conventionally, the width of the hysteresis DB can be identified from bump tests or ramp tests. In the bump test, a series of step changes are made and hysteresis can be observed from the steady state error resulting from opposite direction step changes (Figure 10.6A). Provided with the steady state gain, the DB can be calculated from the steady state error. For systems with a long time constant, the bump test can be time consuming. Another approach is to ramp the control output up first, followed by a downward ramp (Figure 10.6B). The width of the hysteresis can be observed from the plot of y and u_{co}, as shown in Figure 10.6B. For linear systems, the width DB is simply the gap in between. But for a nonlinear system the gap may not be quite as obvious as shown in a later section.

If the width of the hysteresis is available, then we can use the inverse of the hysteresis to adjust the shape of the relay. The inverse of the hysteresis can be expressed analytically.

$$u'_{co}(t) = \frac{u_{co}(t)}{m} + \chi_r(t)c_r + \chi_1(t)c_1 \qquad (10.8)$$

with

$$\chi_r(t) = \begin{cases} 1 & \text{if } u_{co}(t) > u_{co}(t-1) \\ & \text{or } u_{co}(t) = u_{co}(t-1) \text{ and } \chi_r(t-1) = 1 \\ 0 & \text{otherwise} \end{cases} \qquad (10.9)$$

$$\chi_1(t) = \begin{cases} 1 & \text{if } u_{co}(t) < u_{co}(t-1) \\ & \text{or } u_{co}(t) = u_{co}(t-1) \text{ and } \chi_1(t-1) = 1 \\ 0 & \text{otherwise} \end{cases} \qquad (10.10)$$

Figure 10.7 shows corresponding relays if we have the information about the hysteresis. For the ideal relay we simply increase the relay height by a factor of $DB/2$, *i.e.* relay height $= h + (DB/2)$. For the saturation relay the shape of the relay can also be modified to accommodate the hysteresis (Figure 10.7B). Since the inverse of the hysteresis cancelled out the effect of the imperfect actuator, as expected, correct ultimate properties can be obtained using the modified relay (Figure 10.7). Note that the modified relays in Figure 10.7 provide the *exact* counter measures to overcome the hysteresis problems. Despite the fact that the effect of hysteresis can be compensated using the modified relays, we still need two experiments to complete the procedure: finding the width of hysteresis followed by a relay feedback test.

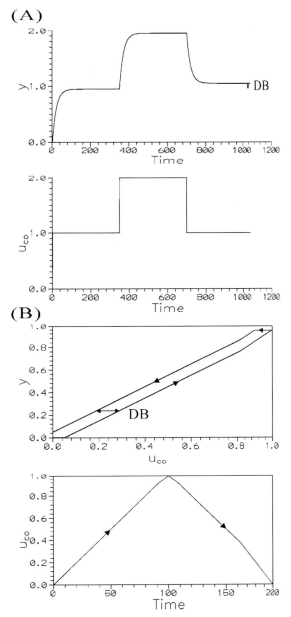

Figure 10.6. Identification of hysteresis using (A) bump test and (B) ramp test

204 Autotuning of PID Controllers

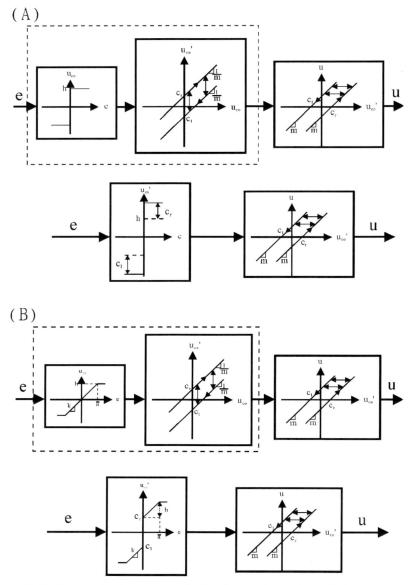

Figure 10.7. Resultant relays by combining the original relay with the inverse of a hysteresis for (A) ideal relay and (B) saturation relay

10.2.2 Simultaneous Identification

Instead of identifying the hysteresis width and ultimate properties separately, it is possible carry out the experiment in one step. As mentioned earlier, the reason the ideal relay overestimates the ultimate gain under hysteresis (with a width DB) is that the actual output of the relay u is reduced by $DB/2$ as the result of a hysteresis. For example, in Figure 10.1A the actual actuator positions are

$$u = h - \frac{DB}{2} \quad \text{or} \quad -\left(h - \frac{DB}{2}\right) \tag{10.11}$$

The amplitude of oscillation a was generated from this relay height. Unfortunately, we still use h to compute the ultimate gain using Equation 10.2. This consequently results in a much larger K_u. This insight leads to a new relay feedback procedure for the simultaneous identification of DB and K_u. If we carry out a relay feedback test using two different relay heights h_1 and h_2, this results in sustained oscillations with two amplitudes a_1 and a_2. As the result of possible hysteresis, the actual relay heights are $h_1 - DB/2$ and $h_2 - DB/2$. From sustained oscillations we have

$$\frac{a_1}{a_2} = \frac{h_1 - \frac{DB}{2}}{h_2 - \frac{DB}{2}} \tag{10.12}$$

Therefore, the width of the hysteresis DB can be calculated from

$$\frac{DB}{2} = \frac{a_2 h_1 - a_1 h_2}{a_2 - a_1} \tag{10.13}$$

Once DB becomes available, the ultimate gain can be computed from

$$K_u = \frac{4\left(h_i - \frac{DB}{2}\right)}{\pi a_i} \tag{10.14}$$

where $i = 1$ or 2. The ultimate period P_u can be read directly from the responses, since it will not be affected by the hysteresis (*e.g.* Figure 10.2). The procedure can be summarized as follows:

(1) Select a relay height h_1 (*e.g.* 3–7%).
(2) Perform a relay feedback test and, when the system reaches sustained oscillation, read off the amplitude of oscillation a_1 and ultimate period P_u.
(3) Increase the relay height h_2 by 2–4% and read off the amplitude of oscillation a_2.
(4) Compute the width of hysteresis DB from Equation 10.13 and the ultimate gain from Equation 10.14.

In step 1, the relay height h_1 should be greater than DB/2 to prevent failure in the experiment, since the actual relay height is discounted by DB/2. In step 3, only a small increase from h_1 (*e.g.* 2–4%), generally, can provide the resolution we need to differentiate a_1 and a_2 (*i.e.* we will have a net effect from the relay height change). Notice that the ultimate periods should be the same for a relay with different heights for linear systems. Any deviation from the equality is an indication of process nonlinearity, non-uniform hysteresis width, *etc.*

10.3 Applications

Since the two-step procedure, in theory, will give a perfect estimation in ultimate properties (provided with an exact value of DB), the second approach is tested here. Let us use two linear systems and one nonlinear process to illustrate the simultaneous identification procedure for systems with imperfect actuators. Comparisons are made between conventional and proposed relay feedback tests.

10.3.1 Linear Systems

10.3.1.1 Noise-free System

Let use a transfer function model to illustrate the identification procedure. First, consider a noise-free linear model.

Example 10.1 Noise-free example
FOPDT system:

$$G(s) = \frac{e^{-s}}{20s+1} \tag{10.15}$$

Suppose the actuator exhibits hysteresis with a width DB = 5%, $m = 1$ and $|c_l| = |c_r|$. An ideal relay feedback with $h = 5\%$ gives $K_u = 52.9$, which is almost 100% larger than the nominal value ($K_u = 26.5$, *e.g.* Equation 10.7). Following the proposed procedure, we start with $h_1 = 4\%$. From process responses (Figure 10.8), we obtain $a_1 = 0.00072$ and $P_u = 3.85$. After a few oscillations the relay height is increased by 3% ($h_2 = 7\%$), as shown in Figure 10.8. The amplitude of oscillation then becomes $a_2 = 0.00217$ and P_u stays the same. The ultimate period indicates that this is a linear system. Then, we proceed to calculate the width of hysteresis using Equation 10.13:

$$\frac{DB}{2} = \frac{0.00217 \times 0.04 - 0.00072 \times 0.07}{0.00217 - 0.00072} = 0.0251$$

This is a very good estimate of DB (*i.e.* 4% error). Once DB is available, the ultimate gain is calculated immediately from Equation 10.14. That gives $K_u = 26.35$. A PI controller is used to control the linear system with an imperfect actuator. The controller is tuned using the modified Ziegler–Nichols method,

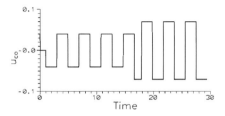
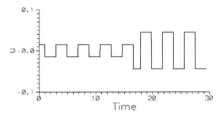

Figure 10.8. Proposed relay feedback tests on an FOPDT system with $h_1 = 4\%$ and $h_2 = 7\%$

$K_c = K_u/3$ and $\tau_I = 2P_u$. Simulation results show that the proposed procedure gives satisfactory responses under an imperfect actuator (Figure 10.9). On the other hand, since the conventional relay feedback does not realize the existence of hysteresis, an oscillatory SP response is observed. ∎

This example clearly indicates that it is possible to identify the width of a hysteresis and ultimate properties simultaneously. Moreover, the estimates appear to be quite accurate compared with the case of a perfect actuator. Any practical identification procedure should be robust with respect to process and/or measurement noises.

10.3.1.2 Systems with Measurement Noise

The proposed method is tested against measurement noise. In the context of system identification, the NSR can be expressed as

$$\text{NSR} = \frac{\text{mean}(\text{abs}(\text{noise}))}{\text{mean}(\text{abs}(\text{signal}))} \tag{10.16}$$

where abs($\cdot$) denotes the absolute value and mean($\cdot$) represents the mean value.

208 Autotuning of PID Controllers

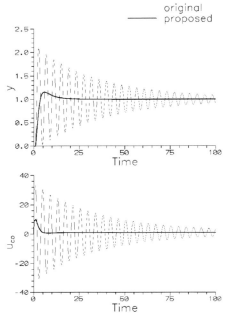

Figure 10.9. SP responses using controller settings from the original and proposed relay feedback tests

Example 10.2 System with measurement noise

Example 10.1 with measurement noise. Consider the case of NSR = 22%. Again, assume the actuator shows hysteresis with a width DB = 5%. First we perform a relay feedback test with $h_1 = 4\%$, followed by a second test with $h_2 = 8\%$ (Figure 10.10). Notice that it is a common practice to employ a relay with hysteresis for systems with significant measurement noise. Here, the width of the hysteresis of the relay is set to twice the standard deviation of the noise. From system output (Figure 10.8), we obtain $a_1 = 0.00081$ and $a_2 = 0.00281$.

The width DB can be computed immediately from Equation 10.13 and, subsequently, the ultimate gain is calculated immediately from Equation 10.14. This results in DB = 4.76% and $K_u = 25.46$. Notice that the estimated ultimate period is 15% higher than the nominal value. Table 10.1 compares the estimates for systems with and without noise. The results show that the proposed method works reasonably well under noisy conditions. As expected, as the noise level decreases (*e.g.* NSR = 10%), a slightly better estimate of DB can be obtained, as shown in Table 10.1. Moreover, SP responses of the noisy system are practically the same as Figure 10.8.

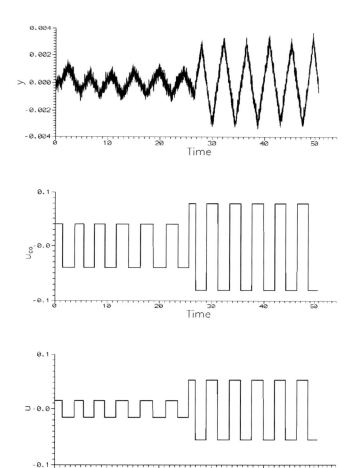

Figure 10.10. Proposed relay feedback tests on an FOPDT system with measurement noise (NSR = 22%)

Table 10.1. Estimates of width of hysteresis and ultimate properties under different NSRs

NSR (%)	$DB_{estimated} / DB_{true}$	$K_u{}^a$	$P_u{}^a$
0	5.02/5.00	26.35	3.85
10	4.86/5.00	26.65	4.07
22	4.76/5.00	25.46	4.32

[a] True valves: $K_u = 26.46$ and $P_u = 3.85$.

210   Autotuning of PID Controllers

Another factor that affects the accuracy in the estimation is the choice of h_1 and h_2. In theory, for linear systems, the results should remain the same regardless of the selection of relay heights. However, the accuracy of estimation may vary for systems with measurement noise. Again, consider the case with a noise level of NSR = 22%. For a range of h_1 and h_2 the simultaneous procedure gives fairly consistent results, as shown in Table 10.2. That is, for the suggested ranges of relay heights, little difference is observed for different choices of h_1 and h_2.   ∎

Example 10.2 clearly demonstrates that the simultaneous identification procedure is quite robust with respect to measurement noise. However, the user should be cau-tious when dealing with measurement noise. First, the amplitude for each oscillation is obtained by taking the average of several points around the peak (seven data points in this case) and the amplitude of oscillation a_i is the average from a few oscillations. The ultimate period P_u is also an averaging value from cycling. Certainly, the noise-handling method should also be applied to a conventional relay feedback test. Moreover, the selection of the relay heights has little impact on the accuracy of the estimation for the linear system studied.

10.3.1.3 Load Disturbance

Low-frequency load change is another important issue that any practical identification procedure has to face. A relay feedback method was proposed to overcome load disturbance when identifying ultimate gain and ultimate frequency. It was proven effective for perfect actuators. Here, the proposed method is extended to handle actuators with hysteresis under load changes.

Let us use the FOPDT system (Example 10.1) to illustrate the identification under load disturbance. Consider the following load transfer function:

$$G_L(s) = \frac{e^{-0.5s}}{10s+1} \qquad (10.17)$$

Table 10.2. Estimates of width of hysteresis and ultimate properties for different relay heights (h_1 and h_2) under 22% NSR

h_1/h_2 (%)	$DB_{estimated}/DB_{true}$	K_u^a	P_u^a
3/6	4.95/5.00	25.79	4.08
3/8	4.92/5.00	25.56	4.58
3/10	4.96/5.00	25.47	4.23
4/6	4.34/5.00	26.50	4.65
4/8	4.76/5.00	25.46	4.32
4/10	4.56/5.00	25.46	4.15

[a] True valves: $K_u = 26.46$ and $P_u = 3.85$.

Again, the actuator shows hysteresis with a width of 5% ($DB = 5\%$). A step load disturbance comes into the system when a relay feedback test with $h = 4\%$ is performed (Figure 10.11) and it gives an unbalanced period of oscillation (e.g. time < 47 in Figure 10.11). We continue the procedure in Section 10.3 with a second relay feedback test ($h = 7\%$) and K_u and P_u can be computed from Equation 10.2 as shown in Figure 10.11. Following the proposed procedure, the width of the hysteresis can be computed from Equation 10.13:

$$\frac{DB}{2} = \frac{0.00215 \times 0.04 - 0.00073 \times 0.07}{0.00215 - 0.00073} = 0.0246$$

This gives $DB = 4.92\%$, which is a fairly good estimation and the ultimate gain is corrected according to Equation 10.14. The corrected ultimate gain becomes $K_u = 26.90$. Notice that, up to this point, we just follow the proposed procedure and biased oscillations are observed as the result of load change (e.g. time < 60 in Figure 10.11). But fairly good estimates in DB, K_u and P_u are obtained as shown in Figure 10.11 (e.g. comparing with the true values in Table 10.1). We can restore a symmetric oscillation using a biased relay, and the biased value δ_0 is

$$\delta_0(s) = -\frac{h\Delta a}{a} \tag{10.18}$$

where Δa is the biased value of the output y. Once the biased relay is installed and output oscillation becomes symmetric, a little better estimation in K_u and P_u can be achieved (Figure 10.11). The result, again, illustrates the effectiveness of the proposed procedure under low-frequency disturbances.

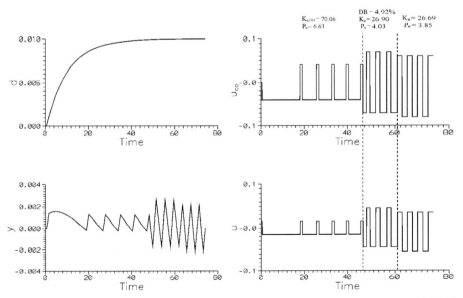

Figure 10.11. Extended relay feedback tests under step load change for an FOPDT system with $h_1 = 4\%$ and $h_2 = 7\%$

10.3.2 Nonlinear Process

The third example is a reactor/separator plant studied by Wu and Yu [9]. The feed to the system is the reactant A and almost high-purity (0.9895 mole fraction) product B is taken out from the bottoms of the distillation column. A conventional control structure is employed (Figure 10.12) and controller settings are obtained using the sequential autotuning approach of Shen and Yu. This is a multivariable system where the product quality x_B is maintained by changing the vapor boil-up V and the distillate composition is controlled by varying the reflux flow R. The nominal production rate is $B = 460$ lb mol/h.

Because of wear, the steam valve exhibits hysteresis with a width of 6%. Not recognizing this fact, controller settings (Table 10.2) from the sequential autotuning simply lead to unacceptable closed-loop responses for a 10% production rate increase, as shown in Figure 10.13.

10.3.2.1 Two-step Procedure

The ramp test (Figure 10.6B) is employed to find the width of the hysteresis DB. For the slow chemical process, it takes an extremely long time (100 h as shown in Figure 10.14) to complete the ramp test while finding a reasonable value for the DB. Following the standard procedure, the width of the hysteresis can be observed from the plot of the controlled variable x_B versus controller output V_{co} (Figure 10.15). This gives a DB of 5.6% (a little lower than the true value of 6%). It is interesting to note that nonlinear characteristics are observed in the ramp test (Figure 10.15) as opposed to the linear system (Figure 10.6B). Once DB is available, we can compensate the offset from the hysteresis by adjusting the relay height (Figure 10.7). After the compensated relay feedback tests, the controller settings immediately become available, as shown in Figure 10.14. Despite being able to correctly identify ultimate properties, the ramp test part of the procedure is simply too time consuming and may not be a good choice for slow chemical processes. On the contrary, the relay feedback type of test becomes attractive.

10.3.2.2 Simultaneous Procedure

The simultaneous approach uses consecutive relay feedback tests to identify the width of the hysteresis and the ultimate properties. Following the Shen–Yu tuning, we can find the controller parameters (Table 10.3). Figure 10.16 shows the sequence of autotuning when the steam valve V (shown in the graph) is imperfect. It takes less than 10 h to complete the tuning of top and bottom loops (as opposed to more than 100 h for the two-step procedure). The new settings give good closed-loop performance (Figure 10.13) which is not too different from that when the steam valve does not have hysteresis. It should be noted that the actual opening of the steam valve V is quite different from the control output V_{co}. It is clear that the behavior of the steam valve movement is far from sustained cycling as the result of hysteresis (Figure 10.16). The situation becomes worse if both the reflux and steam valves exhibit hysteresis. Figure 10.17 shows that a prolonged transient response occurs during the tuning of the x_D loop. A longer experimental time or, even worse, possible failure in the experiment can be expected when more and more valves in the system show hysteresis. Despite the fact that the settings from Figure 10.17 work almost as good as the previous case, the efficiency in the relay feedback test deteriorates quickly when the number of imperfect actuators increases. This implies that, regardless of how smart the autotuner may be, the best solution to handle imperfect actuators is to have the valves fixed or to get a positioner.

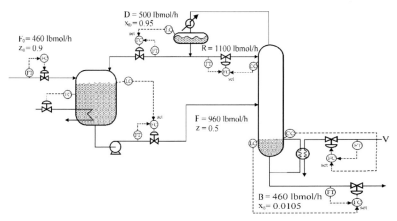

Figure 10.12. Recycle plant with R–V control structure on distillation column

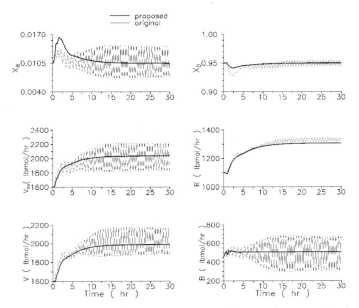

Figure 10.13. Step responses of recycle plant with imperfect actuator in the bottoms loop for 10% production rate increase using controller settings from proposed and original autotuning methods

214 Autotuning of PID Controllers

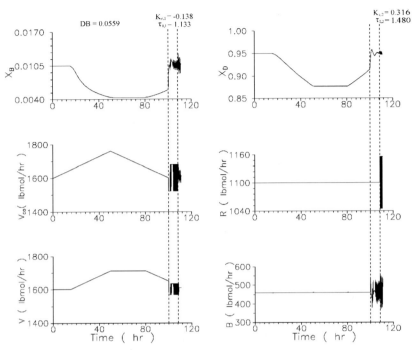

Figure 10.14. Ramp test followed by relay feedback tests for the recycle plant with imperfect actuator in the bottoms loop

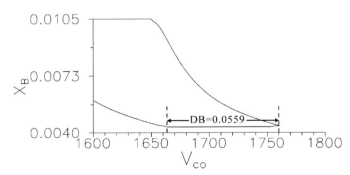

Figure 10.15. Plot of the controlled variables and controller output from ramp test to identify width of the hysteresis

Imperfect Actuators 215

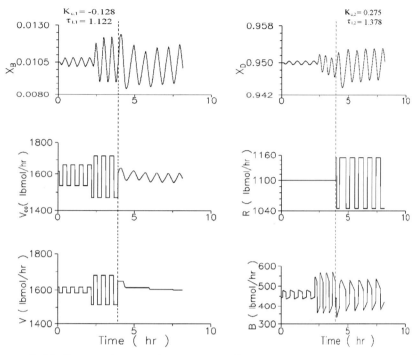

Figure 10.16. Proposed autotuning sequence for a recycle plant with an imperfect actuator in the bottom loop

Table 10.3. Controller parameters using different autotuning methods for the recycle plant under various situations of imperfect valves

	$DB_{estimated} / DB_{true}$ (%)		K_c / τ_I [b]	
	Bottom	Top	x_B [a]	x_D [a]
Original	−/0	−/0	−0.126/1.116	0.275/1.289
Original	6/6	−/0	−0.352/1.101	0.121/1.350
Proposed	6/6	−/0	−0.128/1.122	0.275/1.378
Original	6/6	5.9/6	−0.369/1.138	0.169/1.360
Proposed	6/6	5.9/6	−0.127/1.124	0.272/1.383

[a] Transmitter spans: $x_B = 0.021$ and $x_D = 0.1$
  mole fraction valve gains: twice nominal steady state flow rate.
[b] In h.

216 Autotuning of PID Controllers

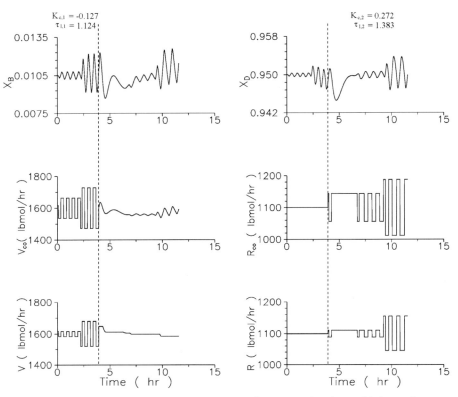

Figure 10.17. Proposed autotuning sequence for a recycle plant with imperfect actuators in both top and bottom loops

10.4 Conclusion

Hysteresis is sometimes observed in pneumatic and piezoelectric actuators. Under relay feedback, the effects of actuator hysteresis on estimation errors are explored. Comparisons are made between the ideal and saturation relays. As expected, the ramp behavior of the saturation relay can alleviate the overestimate of K_u. If the width of hysteresis DB is available (*e.g.* estimated, from bump test, *etc*), the shapes of the relays can be modified to accommodate the imperfection and to provide better accuracy. Moreover, a procedure for simultaneous identification of the width of the hysteresis and ultimate properties is also proposed. Simulation results show that good estimates of ultimate properties can be obtained. It provides better reliability for relay feedback identification and opportunities for improved control under imperfect actuators. Finally, a word of caution is required: regardless of how smart the autotuner is, the best approach to handling an imperfect actuator is to have it fixed.

10.5 References

1. Ender DB. Process control performance: Not as good as you think. Control Eng. 1993;40:180.
2. Lin Z. Robust semi-global stabilization of linear systems with imperfect actuators. Syst. Control Lett. 1997;29:215.
3. Lewis FL, Tim WK, Wang LZ, Li ZX. Deadzone compensation in motion control systems using adaptive fuzzy logic control. IEEE Trans. Control Syst. Tech. 1999;7:731.
4. Tao G, Kokotovic PV. Adaptive control of systems with unknown output backlash. IEEE Trans. Automat. Control 1995;40:326.
5. Kimura T, Hara S, Fujita T, Kagawa T. Feedback linearization for pneumatic actuator systems with static friction. Control Eng. Prac. 1997;5:1385.
6. McMillan GK. Tuning and control loop performance. Instrument Society of America: Research Triangle Park; 1994.
7. Wallen A. Valve diagnostic and automatic tuning. Proc. ACC 1997; 2930.
8. Åström KJ, Hägglund T. PID controllers: Theory, design and tuning. Instrumentation Society of America: Research Triangle Park; 1995.
9. Wu KL, Yu CC. Reactor/separator process with recycle 1. Candidate control structure for operability. Comput. Chem. Eng. 1996;20:1291.

11

Autotuning for Plantwide Control Systems

Typical chemical processes consist of many process units. Therefore, the success of the production depends a great deal on the smooth operation of all these units. As a result of stringent environmental regulation and economic consideration, today's chemical plants tend to be highly integrated and interconnected. Moreover, the steady state and dynamic behavior of these interconnected units differs significantly from individual units. Therefore, the problem of plantwide control becomes the operation and control of these interconnected process units. A typical interconnected process unit is the recycle system: process with material recycle.

The dynamics and control of processes with recycle streams received little attention until recently. Luyben and Tyreus [1–4] investigated the effects of recycle loops on process dynamics, and the interaction between design and control was also studied for several process systems with different levels of complexity, *e.g.* different numbers of process units and chemical species. Downs and Vogel [5], based on a commercial process system, proposed a benchmark plantwide control problem, the Tennessee Eastman problem, for the purpose of developing, studying and evaluating process control technology. Luyben *et al.* [6] give good guidelines and a summary on plantwide control. Now, plantwide control is included in typical process control textbooks [7–9].

As pointed out in Chapter 1, several ways exist to improve control performance: (1) better process design, (2) selecting an appropriate control structure, (3) improved controller settings and (4) better instrumentation. Issues (2) and (3) are addressed in this chapter. First, we will show that control structure design plays a significant role in improving control performance. Once the control structure is fixed, a procedure is proposed for the tuning of the entire plant.

11.1 Recycle Plant

Before looking into process characteristics, a simple reactor/separator process is described. The process studied is a flowsheet consisting of a reactor and a distillation column in an interconnected structure as shown in Figure 11.1 [10]. An

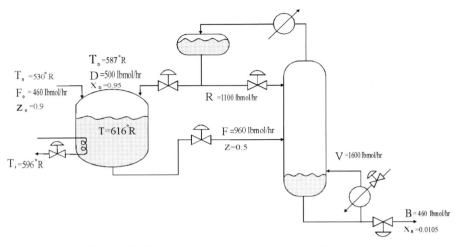

Figure 11.1. A reactor/separator process with recycle

irreversible first–order reaction ($A \rightarrow B$) occurs in a continuous stirred tank reactor (CSTR). The reaction rate k is a function of temperature described by the Arrhenius expression, i.e. $k(T) = k_0 e^{-E/RT}$. This is an exothermic reaction and the reactor temperature T is controlled by manipulating the cooling water flow rate. Some of the reactant is consumed in the CSTR, and the effluent of the reactor, a mixture of A and B, is fed into a 20-tray distillation column. The product is taken out from the bottoms of the column and the purified reactant is recycled back into the CSTR. The column has a partial reboiler and a total condenser. Constant relative volatility ($\alpha = 2.0$) is assumed for the modeling purpose.[1] Table 11.1 gives the nominal operating condition for the process. The steady state equations play an important role in analyzing this recycle system. From material balances, we have

$$\text{reactor}: \quad F_0 + D = F \quad (11.1)$$

$$F_0 z_0 + D x_0 = F z + V_R k z \quad (11.2)$$

$$\text{column}: \quad F = D + B \quad (11.3)$$

$$F z = D x_D + B x_B \quad (11.4)$$

$$\text{overall}: \quad F_0 = B \quad (11.5)$$

Notice that the external flows into and out of the system are the reactor fresh feed flow rate F_0 and the column bottoms flow rate B respectively. Rearranging Eqs 11.2, 11.4 and 11.5 we obtain

---

[1] A FORTRAN program for this plantwide control system can be obtained by contacting the author, e-mail: ccyu@ntu.edu.tw

Table 11.1. Parameter values and steady state condition for reactor/separator system

CSTR		
Fresh feed flow rate (F_0)	460.000	(lbmol/h)
Fresh feed composition (z_0)	0.900	(mole fraction)
Fresh feed temperature (T_0)	530.000	(°R)
Recycle flow rate (D)	500.378	(lbmol/h)
Recycle stream composition (x_D)	0.95	(mole fraction)
Recycle stream temperature (T_D)	587.156	(°R)
Reactor temperature (T)	616.425	(°R)
Coil temperature (T_j)	596.070	(°R)
Reactor holdup (V_R)	2400.945	(lbmol)
Activation energy (E)	30841.770	(Btu/lbmol)
Pre-exponential factor (k_0)	$2.8297 \cdot 10^{10}$	(h^{-1})
Reactor residence time	2.500	(h)
Overall heat-transfer coefficient (U)	150.519	(Btu/h/ft°R)
Heat-transfer area (A)	3206.8	(ft^2°R)
Heat capacity (C_p)	0.750	(Btu/lb$_m$/°R)
Heat of reaction (λ)	−30000.000	(Btu/lbmol)
Density (ρ)	65.35	(lb$_m$/ft^3)
Molecular weight (MW)	60.05	(lb$_m$/lbmol)
Distillation		
Column feed flow rate (F)	960.378	(lbmol/h)
Column feed composition (z)	0.500	(mole fraction)
Reflux flow rate (R)	1100.045	(lbmol/h)
Distillate flow rate (D)	500.378	(lbmol/h)
Reflux ratio (RR)	2.198	(mole fraction)
Bottoms flow rate (B)	460.000	(lbmol/h)
Vapor boil-up (V)	1600.423	(lbmol/h)
Bottoms composition (x_B)	0.0105	(mole fraction)
No. of trays (NT)	20	
Feed tray (NF)	12	
Relative volatility (α)	2	
Liquid hydraulic time constant (β)	0.0011	(h)
Bottoms holdup (M_B)	275	(lbmol)
Reflux drum holdup (M_D)	185	(lbmol)
Tray holdup (M_n)	23.5	(lbmol/tray)

$$F_0(z_0 - x_B) = kV_R z \qquad (11.6)$$

and rearranging Equations 11.3, 11.4, and 11.5 we obtain

$$\frac{F}{F_0} = \left(\frac{x_D - x_B}{x_D - z}\right) \qquad (11.7)$$

Equations 11.6 and 11.7 give an insight into this reactor/separation system. For example, three possible process variables to handle external load changes, *i.e.* changes in F_0 or z_0, are reactor holdup V_R, reaction rate constant k and reactor composition (mole fraction of light component) z. Conventionally, V_R (via level control) and k (via reactor temperature control) are kept constant and this, subsequently, results in a large change in the reactor composition z. Once significant deviation occurs in z, this results in large changes in the internal flows (*e.g.* F, as shown in Equation 11.7). Therefore, the disturbance rejection capabilities of different control structures can be analyzed from these steady state equations (Equations 11.1–11.7). The experience of the disturbance rejection capability of each individual unit can be useful for recycle systems.

11.2 Control Structure Design

For a system with multiple units, alternatives exist for handling load disturbances. For example, the effect of a throughput change can mostly be absorbed by a single unit or it can be evenly handled by all units. Inappropriate disturbance handling can lead to unreasonable demand on the capacity of an individual unit and, consequently, result in the *snowball* effect [11].

11.2.1 Unbalanced Schemes

In a recyce plant, if load changes are handled mostly by one single unit in a plant-wide system, some of the process variables (*e.g.* flow rates, level, *etc.*) can hit operational constraints for a very small load change.

11.2.1.1 Column Overwork

Let us first consider the conventional control structure (Figure 11.2a) where the reactor holdup V_R is kept constant by changing the reactor effluent flow rate F. On the column side, both the top and bottoms compositions (x_D and x_B) are controlled by manipulating the reflux flow rate R and vapor boil-up V respectively. A distinct feature of this structure is that the reactor holdup is kept constant, as shown in Figure 11.2a. This practice gives little problem for plants that are connected as cascade units. However, for recycle systems, the practice of constant reactor holdup may require the separator to work much harder in order to maintain product specifications. Consider the case of a throughput F_0 increase.

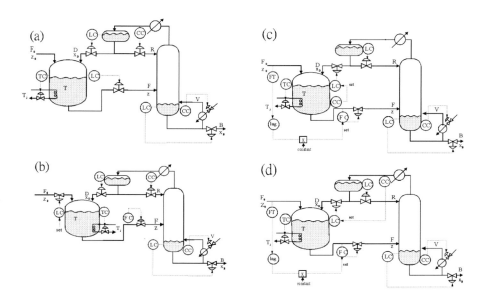

Figure 11.2. Control structures for reactor/separator process: (a) conventional structure (structure C), (b) Luyben structure (structure L), (c) balanced structure 1 (structure B_1) and (d) balanced structure 2 (structure B_2)

For a reactor (as an individual unit), the reactor holdup has to increase proportionally in order to keep the same performance (*e.g.* Equation 11.6). In this conventional structure, since the reactor level is kept constant, the reactor effluent composition z (or the column feed composition) remains high (as the result of a smaller residence time) along with an increased feed flow rate F (Figure 11.3). The increases in both the feed flow rate F and feed composition z make the column boil much more of light component to the top, which it subsequently recycles back to the reactor in order to maintain the product specification. In fact, the resultant process variables can be derived analytically for the fresh feed flow rate F_0 changes. Assuming constant k and V_R, from Equations 11.1–11.4 we have

$$\left(\frac{F}{\overline{F}}\right)_c = \left(\frac{\overline{x}_D - \overline{z}}{\overline{x}_D - r\overline{z}}\right) r \tag{11.8}$$

where the overbar denotes nominal steady state value, $r = F_0 / \overline{F}_0$ and the subscript c denotes the conventional structure. Similarly, the distillate flow rate, reactor composition z and reactor holdup V_R can also be expressed as

$$\left(\frac{D}{\overline{D}}\right)_c = \left[\frac{\overline{x}_D - \overline{z}}{\overline{x}_D - r\overline{z}}\left(1 + \frac{\overline{F}_0}{\overline{D}}\right)\right] r \tag{11.9}$$

$$\left(\frac{z}{\bar{z}}\right)_c = r \qquad (11.10)$$

$$\left(\frac{V_R}{\bar{V}_R}\right)_c = 1 \qquad (11.11)$$

Equation 11.10 clearly shows that the constant level practice of the conventional structure results in an *underperformance* of the reactor, *e.g.* percentage change in z is proportional to percentage change in F_0. This, subsequently, requires overwork in the column. First, one can observe an *ultimate constraint* imposed on this structure from Equations 11.8 and 11.9. If we have

$$r = \frac{\bar{x}_D}{\bar{z}} = 1.9 \qquad (11.12)$$

the distillate flow rate (or the column feed flow rate) goes to infinity, as can be seen from Equation 11.9. Obviously, any process variable has a physical constraint, *e.g.* the maximum flow capacity in the distillate is often designed as twice $\bar{D}$. This means that the operability (throughput handling ability) of the conventional structure is much smaller than the data from Equation 11.12 (Figure 11.3). This is exactly the snowball effect pointed out by Luyben [11]. In order to maintain the desired separation under an increase in the production rate, the distillation column has to handle increases in both the feed composition z and feed flow rate F. Therefore, both the vapor boil-up V and reflux flow rate R increase quadratically for a linear increase in F_0, as shown in Figure 11.3. Since only a fixed amount of product ($B = F_0$) is taken out of the column, most of these flow rate increases recycle back to the reactor. Figure 11.3 shows the changes in the process variables for a range of changes in $F_0/\bar{F}_0$ (from 0.1 to 1.6). The process behavior shown here is very different from that of cascade units or of individual units. Furthermore, this result comes from an almost unnoticed reason that the reactor does not maintain its performance during a throughput change. Actually, to some extent, Equation 11.6 does reveal this fact. For a given product specification x_B, the load changes in F_0 and z_0 can only be handled via k, V_R or z. A constant holdup V_R combined with a constant reactor temperature control strategy (a common practice for cascade units) results in a column overwork (*i.e.* column feed composition z absorbing all the changes in load variables).

Similar behavior can also be observed for changes in the fresh feed composition. Figure 11.4 shows how process variables vary for a range of z_0 changes. Wu and Yu [12] give an analytical expression for relationships between process variables.

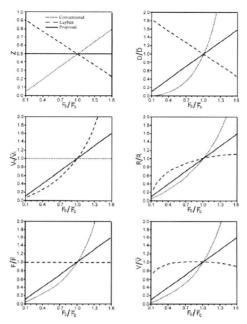

Figure 11.3. Steady state values of process variables for a range of F_0 changes under different control structures

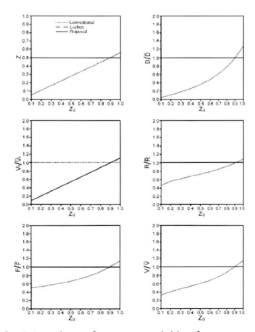

Figure 11.4. Steady state values of process variables for a range of z_0 changes under different control structures

11.2.1.2 Reactor Overwork

Luyben [11] recognized the effect of column overwork and the potential problem of the snowball effect on the recycle stream. A new control structure is proposed (Figure 11.2b). In the Luyben structure, the reactor holdup is adjusted for known changes in fresh feed flow F_0 and feed composition z_0. This, in fact, overcomes the reactor underperformance problem. However, a unique feature of the Luyben structure is that the reactor effluent flow rate F is kept constant using a flow controller (Figure 11.2b). This implies that, even under throughput changes, the column feed flow rate is not allowed to change. As for the column control, both the top and bottoms compositions are controlled by manipulating R and V respectively, as shown in Figure 11.2b. Despite the fact that the fresh feed flow is used as a manipulated variable in the Luyben structure, the throughput change is accomplished in an indirect manner, i.e. by adjusting the reactor level. Again, the process variables can be expressed analytically by solving Equations 11.1–11.4. The variables kept constant are x_D, x_B and F. For throughput changes ($r = F_0 / \overline{F_0}$), the corresponding process variables are

$$\left(\frac{F}{\overline{F}}\right)_L = 1 \tag{11.13}$$

$$\left(\frac{D}{\overline{D}}\right)_L = \left(1 + \frac{\overline{F_0}}{\overline{D}}\right) - \left(\frac{\overline{F_0}}{\overline{D}}\right) r \tag{11.14}$$

$$\left(\frac{z}{\overline{z}}\right)_L = \frac{\overline{x_D}}{\overline{z}} - \left(\frac{\overline{x_D}}{\overline{z}} - 1\right) r \tag{11.15}$$

$$\left(\frac{V_R}{\overline{V_R}}\right)_L = \frac{\overline{z}}{\overline{x_D} - (\overline{x_D} - \overline{z}) r} r \tag{11.16}$$

where the subscript L denotes the Luyben structure. Figure 11.3 shows the changes in the process variables for a range of throughput changes ($F_0 / \overline{F_0} = 0.1 - 1.6$). The results clearly show that the variable reactor holdup structure does alleviate the snowball effect on the recycle stream (e.g. $D / \overline{D}$ in Figure 11.3). However, in this variable-reactor-level control structure, an important question to ask is what an appropriate reactor holdup is. For an individual reactor, the reactor performance is maintained by keeping reactor composition z constant. Since F is kept constant, an increase in V_R (as a result of an increase in the throughput) leads to a larger residence time (V_R / F) and, subsequently, results in a better conversion (a smaller z). Comparing this with the conventional structure (Figure 11.3), the reactor composition z is overadjusted and, subsequently, the process variables in the column remain fairly constant for throughput changes. For this reactor overwork condition, the snowball effect, in fact, remains. Instead of large changes in the recycle stream, the reactor holdup V_R changes significantly for a throughput increase (Figure 11.3). The ultimate constraint imposed on the Luyben structure is V_R, as shown in Equation 11.16. If

$$r = \frac{\bar{x}_D}{\bar{x}_D - \bar{z}} = 2.09 \tag{11.17}$$

the reactor holdup goes to infinity. Obviously, in practice, the throughput handling ability is much smaller then this value (*e.g.* a finite capacity imposed on the reactor holdup). Therefore, it becomes obvious that the snowball effect does not disappear for throughput changes. It appears in the reactor holdup V_R instead of in the recycle flow rate.

In the Luyben structure, the fresh feed composition disturbance can be handled by adjusting the reactor holdup (Figure 11.4). Moreover, the process variables z, D, F, R, V remain unchanged for such disturbances.

11.2.2 Balanced Scheme

From the analyses of the conventional structure and the Luyben structure, it becomes clear that if the load disturbance is not handled evenly by these two units, then this imbalance grows exponentially via the recycle structure. This, consequently, leads to the snowball effect and, more importantly, results in a limited disturbance rejection capability. That is a unique feature of plantwide control. Therefore, care has to be taken in devising the control structure by distributing extra work evenly between the two process units.

For reactor control, a measure of performance is the reactor composition. The reactor composition can be controlled by adjusting reactor holdup V_R (Figure 11.2c). In doing this, the reactor level grows linearly for changes in fresh feed flow rate, as indicated by Eqs 11.6 and 11.7. As for the distillation column control, since both the column feed flow rate F and composition z are controlled (in a feedforward or feedback manner) for external load changes, only single-end composition control in the separator is sufficient to hold top and bottoms compositions. Once the reactor/separator is controlled in this way, the separator shares its work under a throughput change. It is worthwhile mentioning that, in this structure, the reactor and separator are treated as a complete process unit and the control system is designed accordingly. For example, the recycle flow D is adjusted by measuring the reactor level (Figure 11.2c). This design concept indicates an important point in plantwide control: treat the whole plant as a unit instead of designing a control system for each individual unit and then putting them together to form a plantwide control structure. For this control structure, an analytical expression for process variables under throughput changes can also be derived from Equations 11.1–11.4. By assuming constant z and F/F_0 ratio, the process variables of interest become

$$\left(\frac{F}{\bar{F}}\right)_b = r \tag{11.18}$$

$$\left(\frac{D}{\overline{D}}\right)_b = r \qquad (11.19)$$

$$\left(\frac{z}{\overline{z}}\right)_b = 1 \qquad (11.20)$$

$$\left(\frac{V_R}{\overline{V_R}}\right)_b = r \qquad (11.21)$$

where r is the relative change in the fresh feed flow rate ($F_0/\overline{F}_0$) and the subscript b denotes the balanced control structure. It immediately becomes clear that, comparing this with the other two structures, ultimate constraint does not exist in this structure; this gives better operability. Comparison is made for these three control structures under throughput changes. The results (Figure 11.3) clearly indicate that, for the balanced control structure, the *extensive* variables (*e.g.* V_R, F, V, R) change in proportion to throughput F_0 changes. In other words, both units share their work to overcome throughput changes. On the other hand, if one of the units overworks, the manipulated variables (or process variables; *e.g.* D for conventional structure or V_R for Luyben structure) could be saturated for a small range of load changes. Figure 11.4 shows how these three control structures handle fresh feed composition changes. For both the Luyben and the balanced structures, z_0 changes are handled by the reactor and these two structures show identical results. Wu and Yu [12] give the derivation for fresh feed composition changes.

Notice that the configuration shown in Figure 11.2c (structure B_1) is not the only possible choice to achieve this balance in plantwide control. An alternative is to keep the distillation top composition constant by changing the reactor level SP, as shown in Figure 11.2d (structure B_2). This control structure gives exactly the same disturbance rejection capability as the other balanced structure (Figure 11.3). Equation 11.7 clearly shows that as long as the ratio F/F_0 is kept constant, keeping any two compositions (out of x_D, x_B and z) constant will maintain the third composition at its SP. Therefore, the more appropriate control structure can be selected from these two alternatives according to their dynamic properties. Notice that two composition analyzers are required for all control structures mentioned (Figure 11.2).

11.2.3 Controllability

The RGA of Bristol [13] was employed to analyze the interaction [10] and to assess the controllability of plantwide control systems. It is well known that the RGA is an interaction measure for multivariable systems and it can be used to test the integral controllability of a closed-loop system [14,15]. There are three major loops in this plantwide structure (two composition loops, x_B and x_D or x_D and z, and one temperature loop, T), 3×3 RGAs can be obtained for these four control structures (Figure 11.5) from steady state rating programs (Table 11.2). The results show that all these four structures (structures C, L, B_1 and B_2) are decentralized integral controllable [15]. That is, the controller gains for any of these loops can be reduced arbitrarily to zero (manual mode) without causing instability. Therefore, all three control structures are failure tolerant. This guarantees the integrity of the control system. Furthermore, the closed-loop interaction can also be analyzed using the RGA. Since the temperature loop is much faster than the composition loops, it is easier to interpret the interaction by looking at the composition loops (assuming constant T). Table 11.2 gives RGAs for the reduced systems. Before looking at the plantwide system, it should be noticed that the RGA for the column itself (under $R-V$ control; Figure 11.2) is

$$\Lambda = \begin{matrix} x_B & x_D \\ \begin{bmatrix} 6.8 & -5.8 \\ -5.8 & 6.8 \end{bmatrix} & \begin{matrix} V \\ R \end{matrix} \end{matrix} \tag{11.22}$$

Obviously, Table 11.2 shows that the relative gain λ_{ij} for Luyben's structure ($\lambda_{11} = 12.16$) is much larger than that of the column alone ($\lambda_{11} = 6.8$) or the conventional structure ($\lambda_{11} = 2.78$). That means if the steady state interaction is the only indication of controllability, then the conventional structure is a better choice.

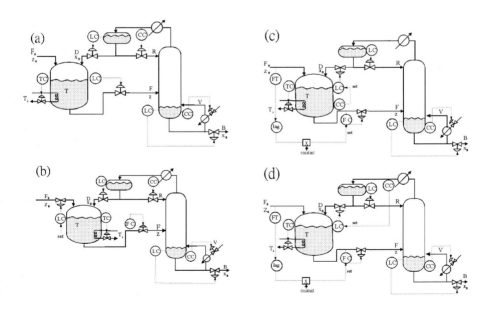

Figure 11.5. Control structures for reactor/separator process with x_B controlled: (a) D fixed (($C)_D$ structure) (b) R fixed (($C)_R$ structure), (c) RR fixed (($C)_{RR}$ structure), and (d) F fixed (($C)_F$ structure)

Table 11.2. RGA for different control structures with different assumptions

Scheme	Controlled variable	Manipulated variable			Manipulated variable*	
		V	R	T_j	V	R
C	x_B	3.303759	−2.078648	−0.225111	2.78	−1.78
	x_D	−2.032999	3.002975	0.030024	−1.78	2.78
	T	−0.270760	0.075672	1.195088		
$(C)_D$	x_B	1.246383	−0.246383		1.0	
	T	−0.246383	1.246383			
$(C)_R$	x_B	1.184476	−0.184476		1.0	
	T	−0.184476	1.184476			
$(C)_{RR}$	x_B	1.207454	−0.207454		1.0	
	T	−0.207454	1.207454			
$(C)_F$	x_B	1.246386	−0.246386		1.0	
	T	−0.246386	1/246386			
L	x_B	9.15964	−8.12565	−0.03399	12.16	−11.16
	x_D	−8.216615	8.916006	0.300144	−11.16	12.16
	T	0.005606	0.209647	0.733847		
		V	V_R^{set}	T_j	V	V_R^{set}
B_1	x_B	0.968	0.147	−0.114	0.78	0.22
	z	0.032	0.512	0.455	0.22	0.78
	T	−0.000002	0.341	0.659		
B_2	x_B	0.731823	0.271457	−0.000328	0.59	0.41
	x_D	0.268179	0.387579	0.344242	0.41	0.59
	T	−0.0002	0.340964	0.6590		

* Assuming constant reactor temperature.

Table 11.2 also shows that the balanced structure has very different characteristics, i.e. $\lambda_{11} = 0.78 < 1$. The RGA for this structure looks very much like a D–V (distillate and vapor boil-up) controlled system. This is quite the case. For the control structure B_2, consider the case when a step increase in V is made. Since D is manipulated by the reactor holdup, the reflux flow increases while keeping D constant. Therefore, the steady state gains for these two compositions have different signs for a change in V. The result is different from the conventional or Luyben control structure, which shows the behavior of the R–V control structure.

The steady state gain matrix is

$$\begin{bmatrix} x_B \\ x_D \end{bmatrix} = \begin{bmatrix} -1.4 \times 10^{-5} & -8.3 \times 10^{-5} \\ 2.3 \times 10^{-5} & -1.6 \times 10^{-5} \end{bmatrix} \begin{bmatrix} V \\ V_R^{set} \end{bmatrix} \quad (11.23)$$

The second column of the gain matrix looks more like the steady state gains for a feed composition change, for a change in V_R results in a change in z and subsequently affects x_D, x_B and z. The D–V structure has a larger closed-loop gain and, therefore, λ_{11} is smaller than unity. The RGA analyses indicate that the input and output pairing is correct for the balanced structure. Furthermore, one can obtain control structures without any interaction by controlling only one-end (bottoms composition). Figure 11.5 shows four possible structures in which only x_B is controlled and x_D is left uncontrolled by fixing one flow rate or flow ratio (e.g. D, R, RR or F). From the interaction point of view, these four structures are better choices (Table 11.2), since the relative gain is unity for SISO systems. If the interaction is the sole measure of controllability, then the least interacting control structure, e.g. single-end control or the conventional structure, should be the candidate control structure. However, the disturbance handling capability seems to be a more important factor in plantwide control structure selection.

11.2.4 Operability

For the control structures studied (Table 11.3), the effects for a range of load changes (z_0 and F_0) can be calculated from the steady state equations. Notice that, in the computation, no constraint is placed on the flow rates or levels. Therefore, the range of load changes that can be handled by the control structure (rangeability) comes from the fact that the product specification(s) (e.g. x_B or x_D and x_B) simply cannot be met. Table 11.3 gives the rangeabilities for all these seven control structures. It is interesting to note that some of the structures give unreasonably small rangeabilities, e.g. $(C)_D$ and $(C)_F$, for fresh feed flow changes. For example, the structure $(C)_D$ can handle only a 3% throughput increase, despite the fact this structure does not have any interaction problem. The reason is that for a positive change in the throughput, the reactor composition z changes accordingly (e.g. Equation 11.10). Therefore, the total light component that goes into the column increases quadratically, which cannot be handled by the column if both x_B and x_D are fixed. That is, the purity of the light component on the top of the column reaches 100% for a 3% increase in the production rate. A similar limitation is observed in the structure $(C)_F$. For an increase in F_0, the distillate flow rate has to be reduced for the fixed reaction effluent flow rate F configuration (Figure 11.5d). Despite the fact that F is flow controlled, the total light component zF

Table 11.3. Disturbance sensitivity analyses for different control structures with different load changes

Scheme	Disturbance variable	$(z_0)_{max}$	$(z_0)_{min}$	$\left(\dfrac{F_0}{\overline{F_0}}\right)_{max}$	$\left(\dfrac{F_0}{\overline{F_0}}\right)_{min}$
Conventional	C	1.0	0.153	1.801	0.131
	$(C)_D$	0.938	0.2898	1.03	0.64
	$(C)_R$	1.0	0.163	1.9	0.2136
	$(C)_{RR}$	1.0	0.162	1.99	0.3
	$(C)_F$	0.945	0.207	1.03	0.67
Luyben	L	1.0	0.189	2.08	0.08
Balanced	(B_1 and B_2)	1.0	0.189	15.12	0.0047

which goes into the column increases as the result of increased purity in x_D. This structure can only tolerate a small increase in F_0 (decrease in D), since it is limited by its physical constrains ($x_D \leq 1$). This can be shown by rearranging Equations 11.5 and 11.6. Denote r as the dimensionless ratio of the fresh feed flow rate to the nominal feed flow rate. We have

$$r = \frac{\overline{F} x_D}{\overline{F}_0 (x_D - \overline{x}_B) + \overline{F}\overline{z}} \qquad (11.24)$$

By substituting the nominal steady state value for $\overline{F} = 960.378$, $\overline{F}_0 = 460$, $\overline{x}_B = 0.0105$, and $\overline{z} = 0.5$ and the limiting value $x_D = 1$ into Equation 10.24, one obtains

$$r = 1.03 \qquad (11.25)$$

Obviously, this shows a complete lack of operability in plantwide control. Unfortunately, the interaction analysis does not give any indication of a limited rangeability.

Again, the balanced schemes give the largest rangeability for throughput changes, as shown in Table 11.3. Furthermore, the Luyben structure has a larger rangeability then the conventional scheme for changes. The results presented here are in contradiction with those from interaction analyses. Therefore, a trade-off has to be made between interaction and operability. All the control structures with x_B and x_D controlled handle z_0 changes equally well. From the ongoing analyses, it becomes obvious that the balanced structure is a better choice from a steady state point of view.

11.3 Controller Tuning for Entire Plant

The dynamics of the reactor/separator process are analyzed using a series of rigorous dynamic simulations. The reactor is a CSTR with the reactor temperature controlled by the cooling water flow rate. The assumptions of theoretical tray, equimolar overflow and constant relative volatility are made in modeling the distillation column. The differential equations are similar to those of Luyben [16] pp. 64, 70. Parameters characterizing dynamic behavior, e.g. holdups in column and reactor, are given in Table 11.1. Constraints are placed on the flow rates and levels. The maximum flow rate and holdup are set to be twice the nominal steady state values except for the fresh feed flow rate, which was set to be three times the steady state value. An analyzer dead time of 6 min and a temperature measurement lag of 1 min are assumed in the composition loop and temperature loop respectively.

11.3.1 Tuning Steps

Despite the fact that many methods have been proposed for the tuning of multivariable systems [16,17], little is said about the tuning of the plantwide control structure in a systematic manner. Several authors [1,18] find ultimate gain K_u and ultimate frequency ω_u first, followed by the Ziegler–Nichols type of tuning method in their plantwide control systems. The Tyreus–Luyben tuning (Table 2.3) is an alternative in plantwide control. Since, typically, many loops are involved in a plantwide system, an important question to be answered is which loop (or group of loops) should be tuned first and by what method. That is, what is the tuning sequence (*e.g.* arranged by unit, by properties or by speed of response)? One thing is clear, however, the inventory loops should be under control when the quality loops are tuned [18].

11.3.1.1 Inventory Control

In this work, the inventory in the system is maintained through three level loops (Figure 11.2). The level loops are tuned first, followed by finding the tuning constants for the composition and temperature loops. Since the holdups in the column (M_D and M_B) are an order of magnitude smaller than that of the reactor holdup V_R (Table 11.1), perfect level control is assumed in these two level loops (controlling M_D and M_B).

The averaging level control of Cheung and Luyben [19] is used for the tuning of the reactor level loop. For the conventional and Luyben structures, a PI controller is employed for the reactor level control. First, the closed-loop time constant is set to be a ratio (*e.g.* 10%) of the reactor residence time and a specific damping ratio ($\zeta = 0.707$) is specified for the closed-loop characteristic equation. Following the tuning procedure of Cheung and Luyben [19], the controller gain K_c and reset time τ_I can be found directly. The tuning constants for the level loops are given in Table 11.4. For the balanced control structures, the reactor level is cascaded by the top composition, and, therefore, a P controller is sufficient to maintain the composition SP. Since a P-only controller is employed in the reactor level control for the balanced structure, the tuning constant is found by setting the closed-loop time constant to be a ratio (roughly 3%) of the residence time (Table 11.4). It should be emphasized that the tuning of the reactor level loop can affect the tuning constants of the quality loops, especially for the Luyben and the balanced structures. The reason is quite obvious: these two structures manipulate the reactor level for quality control.

11.3.1.2 Ratio Control

Since a ratio control is involved in the two balanced control structures (Figure 11.2 c, d), a dynamic element is placed in the feed-forward path. This is a "lag" device with the time constant set to be 10% of the reactor residence time.

234 Autotuning of PID Controllers

Table 11.4. Ultimate properties and controller parameters for different control structures

Structure	Parameter pairing	K_u	ω_u (rad/min)	K_c^*	τ_I (min)
Conventional	Temp. loop $T-T_j$	17.081	94.06	5.69	8.0
	Comp. loop $x_B - V$	−2.24	13.0982	−0.75	57.6
	Comp. loop $x_D - R$	0.92	9.827	0.31	76.7
	Level loop			−5.66	21.2
Luyben	Temp. loop $T-T_j$	17.051	94.2	5.68	8.0
	Comp. loop $x_B - V$	−1.60	9.83	−0.53	76.7
	Comp. loop $x_D - R$	0.81	8.74	0.27	86.4
	Level loop			9.43	44.3
Balanced 1	Temp. loop $T-T_j$	17.064	94.2	5.69	8.0
	Comp. loop $x_B - V$	−9.89	19.592	−3.3	38.4
	Comp. loop $z - V_R^{set}$	−0.16	2.4962	−0.07	125.9
	Level loop			40.29	
Balanced 2	Temp. loop $T-T_j$	16.996	93.919	5.67	8.0
	Comp. loop $x_B - V$	−9.81	20.061	−3.27	37.6
	Comp. loop $z - V_R^{set}$	−0.06	6.874	−0.03	45.7
	Level loop			29.51	

* Transmitter spans: x_D and x_B: 0.1 mole fraction; z: 0.2 mole fraction level: twice nominal steady state holdup.
Valve gains: twice nominal steady state flow rate except for fresh feed flow (three times nominal steady state flow rate).

Furthermore, the dynamic behavior of the two balanced structures (B_1 and B_2) is not quite the same. Consider the case when a step change is made in V_R^{set}. Figure 11.6 shows the responses for the control structure B_1. It is clear that the reactor composition z goes through an inverse response while x_D is showing a little undershoot step response. It is well known that the non-minimum phase behavior of z/V_R^{set} cannot be removed via feedback control. Therefore, the balanced control structure B_2 is selected from the dynamic response point of view and it is used for subsequent comparisons (with other control structures).

11.3.1.3 Quality Loop

Once the inventory is under control, the tuning constants for the reactor temperature and distillation composition loops can be found. PI controllers are employed for quality control. These three loops are tuned using the multivariable autotuner of Chapter 6. The relay feedback MIMO autotuner proceeds with the tuning sequentially and the sequence is repeated until the corresponding tuning constants are relatively close between sequences.

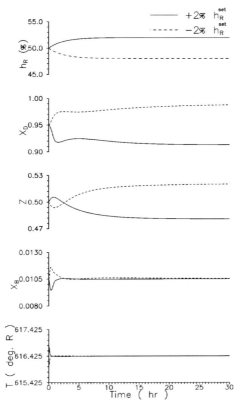

Figure 11.6. Step responses of the control structure B_1 for a step change in reactor level

236 Autotuning of PID Controllers

Let us take the tuning of the conventional structure as an example (Figure 11.7). The sequential tuning approach of Chapter 6 is employed here. Initially, the relay feedback test is performed on the reactor temperature loop and a sustained oscillation is generated as shown in Figure 11.7. The ultimate gain can be found from system responses.

$$K_u = \frac{4h}{\pi a} \qquad (11.26)$$

where a is the amplitude of the output and h is the relay height. The ultimate period P_u can be read off from system responses. Once K_u and P_u are available, K_u and τ_I can be found according to

$$K_c = \frac{K_u}{3} \qquad (11.27)$$

$$\tau_I = 2P_u \qquad (11.28)$$

This gives $K_c = 5.68$ and $\tau_I = 7.8$. Next, the $x_B - V$ loop is under the relay feedback test while the loop is on automatic. The results are $K_c = -0.82$ and $\tau_I = 55.5$. The $x_D - R$ loop is then tuned while the other two loops on automatic.

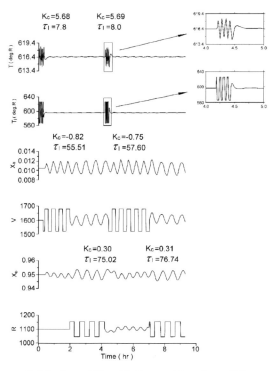

Figure 11.7. Sequential tuning of the $T-T_j$, x_B-V and x_D-R loops for the conventional structure

The tuning parameters for the $x_D - R$ loop are: $K_c = 0.30$ and $\tau_I = 75.0$. Actually, the tuning process can be terminated at this point (over a 4 h period). Figure 11.7 shows that this procedure is repeated for another sequence to ensure that these parameters really converge. Table 11.4 gives the tuning constants for the conventional structure. Following the same procedure, the tuning constants for the Luyben structure can also be found sequentially, as shown in Figure 11.8. The dynamics of these two structures are quite similar (*e.g.* in terms of time required for autotuning as shown in Table 11.4). The balanced structure shows a slightly different characteristic (Figure 11.9). The loop speeds for the $T - T_j$ and $x_B - V$ loops are quite similar to the two structures shown previously. However, the relay feedback test on the $x_D - h_R^{set}$ (level SP) loop takes a much longer time. Despite the fact that the tuning constants converge in one sequence, it takes almost 7 h for one sequence. That implies the loop is much slower than the loops for the other two structures ($x_D - R$ loop). Table 11.4 presents the tuning constants for all three control structures.

In fact, one can learn the dynamic characteristics of the plantwide system from relay feedback tests. For all these structures, the $T - T_j$ loop is much faster than the other two composition loops (almost an order of magnitude faster; Table 11.4).

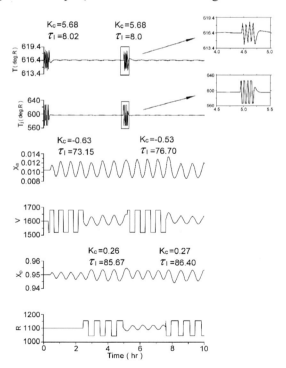

Figure 11.8. Sequential tuning of the $T-T_j$, x_B-V and x_D-R loops for the Luyben structure

238   Autotuning of PID Controllers

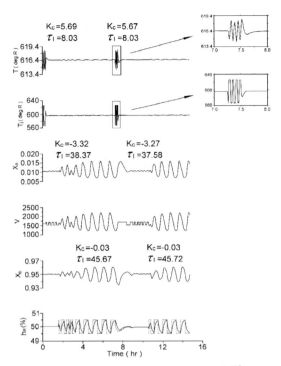

Figure 11.9. Sequential tuning of the T–T_j, x_B–V and x_D–h_R^{set} loops for the balanced control structure B_2

Therefore, the reactor temperature can be treated independently. The next faster loop is the $x_B - V$ loop, as can be seen from the values of ω_u (Table 11.4). For the conventional and Luyben structures the loop speeds for the two composition loops are quite similar, and the difference (in the loop speed) becomes noteworthy for the balanced scheme. The balanced control structure shows a quite different loop speed and gives little dynamic interaction. This can be understood from the fact that the tuning constants from the first sequence and the second sequence are almost the same (Figure 11.9). The autotuning results clearly indicate that the plantwide control structure can be tuned effectively using the sequential tuning approach of Chapter 6.

11.3.2 Closed-loop Performance

The three control structures are tested for the reactor/separator process by performing a series of nonlinear dynamic simulations. Closed-loop performance and operability are employed to measure the effectiveness of these alternative control structures. As far as the product quality is concerned, the response of x_B is the most important indicator among these three controlled variables (x_B, x_D and T).

Figures 11.10–11.12 show what happens when step changes ($\pm 10\%$) are made in the fresh feed flow rate for these three control structures. For the conventional structure, small changes in F_0 ($\pm 10\%$) are amplified into very large deviations in the distillate flow rate ($\pm 30\%$; Figure 11.10). This is exactly the *snowball* effect pointed out by Luyben [11]. Despite the sensitivity problem in the recycle stream, the closed-loop responses, e.g. x_B are reasonably fast. That implies, at least, that the controller settings are satisfactory. Figure 11.11 shows the closed-loop responses for the Luyben structure for $\pm 10\%$ fresh feed flow rate changes. Small changes in F_0 ($\pm 10\%$) result in significant changes in the reactor level ($\pm 20.4\%$). It is also interesting to note that, unlike the conventional structure, the reactor brings the effluent composition z down to 0.452 for a 10% F_0 increase. This is achieved at the expense of a significant increase in the reactor holdup. Figure 11.11 also reveals that the fresh feed flow rate F_0 saturates momentarily when the throughput changes are made. However, the closed-loop performance of x_B is much better than that of the conventional structure (Figures 11.10 and 11.11) despite the sensitivity problem in the reactor level. Figure 11.12 shows what happens when $\pm 10\%$ step changes in the fresh feed flow rate are made for the balanced structure (structure B_2). The results show that the sensitivity problem in either the recycle stream or in the reactor level, observed in the other two structures, no longer exists. The distillate flow rate and reactor level increase in proportion to the increase in the fresh feed flow rate. Figure 11.12 also confirms the finding that $x_D - V_R^{set}$ is the slowest loop in this system. The closed-loop performance of x_B is similar to that of the Luyben structure. Despite the differences in sensitivity and performance of x_B, the closed-loop responses are reasonably fast for all three structures.

A more realistic test for a production rate change, a 30% increase in fresh feed flow rate, is used to evaluate these three control structures. Figure 11.13 shows the closed-loop responses for the conventional, Luyben and the balanced structures. For a 30% increase in F_0 the conventional structure fails to meet the product specifications x_B as the result of control valve saturation in the recycle stream (D in Figure 11.13). For this throughput change, the reactor level almost overflows (levels off at 90%). This occurs despite the fact that x_B is controlled reasonably well. Figure 11.13 shows that, for the balanced structure, good closed-loop performance is achieved without violating (or almost violating) process constraints. Figure 11.14 shows the fresh feed composition decreases by 10%. Again, better responses can be achieved using the balanced structures. Despite the fact the Luyben and the balanced structures show exactly the same steady state behavior, the dynamic behavior differs between loops. The responses of x_D and z are better controlled using the Luyben structure. The conventional structure shows the largest deviation in x_B while the reactor temperature and x_D are under relatively good control. In summary, the balanced control structure gives good closed-loop performance and, more importantly, can handle large load changes without violating process constraints.

240 Autotuning of PID Controllers

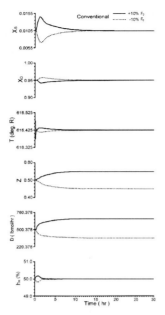

Figure 11.10. Step responses of the conventional structure for ±10% F_0 changes

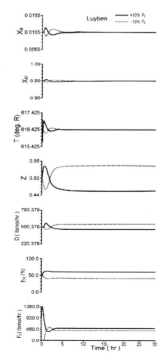

Figure 11.11. Step responses of the Luyben structure for ±10% F_0 changes

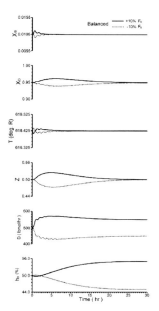

Figure 11.12. Step responses of the balanced (B_2) structure for ±10% F_0 changes

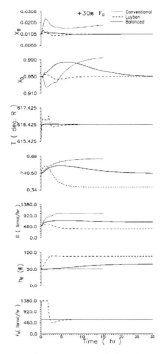

Figure 11.13. Step responses of conventional, Luyben and balanced control structures for +30% F_0 change

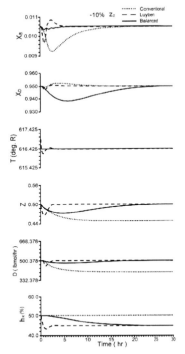

Figure 11.14. Step responses of conventional, Luyben and balanced control structures for -10% z_0 change

11.4 Conclusion

We have provided a thorough study of a simple recycle plant. Two important issues, control structure design and plantwide tuning, are discussed. Both aspects provide opportunities to improve control performance. We have shown that steady state analyses provide useful information to validate the operability of various control structures. The MIMO autotuning procedure of Chapter 6 provides a systematic framework for plantwide control. Simulation results show that the control system is effective in handling large load changes while maintaining good closed-loop performance. Extension to an even simpler control structure can be found in Wu *et al.* [20]. Plantwide control of energy-integrated recycle plant can be found in Lin and Yu [21].

11.5 References

1. Luyben WL. Dynamics and control of recycle systems. 1. Simple open-loop and closed-loop systems. Ind. Eng. Chem. Res. 1993;32:466.

2. Luyben WL. Dynamics and control of recycle systems. 2. Comparison of alternative process designs. Ind. Eng. Chem. Res. 1993;32:476.

3. Luyben WL. Dynamics and control of recycle systems. 3. Alternative process designs in a ternary system. Ind. Eng. Chem. Res. 1993;32:1142.

4. Tyreus BD, Luyben WL. Dynamics and control of recycle systems. 4. Ternary systems with one or two recycle streams. Ind. Eng. Chem. Res. 1993;32:1154.

5. Downs JJ, Vogel EF. A plant-wide industrial process control problem. Comput. Chem. Engng. 1993;17:245.

6. Luyben WL, Tyreus BD, Luyben ML. Plantwide process control. New York: McGraw-Hill; 1999.

7. Luyben ML, Luyben WL. Essentials of process control. New York: McGraw-Hill; 1997.

8. Seborg DE, Edgar TF, Mellichamp DA. Process dynamics and control. 2nd ed. New York: John Wiley & Sons; 2004.

9. Bequette BW. Process control: Modeling, design, and simulation. Prentice-Hall: Upper Saddle River; 2003.

10. Papadourakis A, Doherty MF, Douglas JM. Relative gain array for units in plants with recycle. Ind. Eng. Chem. Res. 1987;26:1259.

11. Luyben WL. Snowball effects in reactor/separator processes with recycle. Ind. Eng. Chem. Res. 1994;33:299.

12. Wu KL, Yu CC. Reactor/separator processes with recycle 1. Candidate control structure for operability. Comput. Chem. Eng. 1996;20:1291.

13. Bristol EH. On a new measure of interaction of multivariable process control. IEEE Trans. Autom. Control 1966;AC-11:133.

14. Morari M, Zafiriou E. Robust process control. Prentice-Hall: Englewood Cliffs; 1989.

15. Yu CC, Fan MKH. Decentralized integral controllability and D-stability. Chem. Eng. Sci. 1990;45:3299.

16. Luyben WL. Process modeling, simulation and control for chemical engineers. 2nd ed. New York: McGraw Hill; 1990.

17. Marino-Galarraga M, McAvoy TJ, Marlin TE. Short-cut operability analysis – 2. Estimation of f_i detuning parameter for classical control systems. Ind. Eng. Chem. Res. 1987;26:511.

18. Price RM, Georgakis C. Plantwide regulatory control design procedure using a tiered framework. Ind. Eng. Chem. Res. 1993;32:2693.

19. Cheung TF, Luyben WL. Liquid level control in single tanks and cascade of tanks with proportional only and proportional integral feedback controllers. Ind. Eng. Chem. Fundam. 1979;18:15.

20. Wu KL, Yu CC, Luyben WL, Skogestad S. Reactor/separator processes with recycle 2. Design for composition control. Comput. Chem. Eng. 2003;27:401.
21. Lin SW, Yu CC. Design and control for recycle plants with heat-integrated separators. Chem. Eng. Sci. 2004;59:53.

12
Guidelines for Autotune Procedure

Intelligent control is becoming a common practice in many industrial applications. Åström and coworkers [1–3] summarize the progress in the field. A review on relay feedback can be found in Hang *et al.* [4]. The reason for such a need is fairly obvious: industrial processes are nonlinear and multivariable in nature, measurements are corrupted with noise and frequent load changes occur. That is, we are dealing with not so *normal* operating conditions in daily practice. In terms of autotuning, this implies we have to devise different experiments to handle various circumstances.

Up to this point we have proposed different types of relay which are useful on various occasions. In this chapter we try to identify typical process characteristics and integrate the relays into corresponding situations. The approach can be implemented in a rule-based expert system, which is available in many modern distributed control systems.

12.1 Process Characteristics

First, we will identify process conditions which are better handled with carefully designed relay feedback experiments. These conditions are typical in many industrial processes.

12.1.1 The Shape

As mentioned in Chapter 3, the shape of relay feedback response gives useful information to identify, at least qualitatively, the model structure (*e.g.* first-order, second-order or HO systems; see Figures 3.10, 4.1 and 4.2) and, possibly, the dead time to time constant ratio. This information is crucial for improved control system performance, because no single tuning rule works well for all model structures over the entire range of parameter values (*e.g.* Tables 2.2 and 2.3). So, we strongly believe the performance of an autotuner can be improved significantly by:

246 Autotuning of PID Controllers

(1) Analyzing the shape of relay feedback response.
(2) Identifying the model structure and find corresponding model parameters.
(3) Applying appropriate tuning rules for the given model structure and parameter values.

More importantly, the improvement can be achieved at virtually no cost, no additional relay feedback test, no prolonged plant test, *etc.*

12.1.2 Load Disturbance

As pointed out in Chapter 7, load changes will lead to erroneous results in K_u and ω_u. A measure is defined to evaluate the effect of load disturbance. The bias-to-signal ratio (BSR) is used to assess the degree of load effect:

$$BSR = \frac{\Delta a}{a} \qquad (12.1)$$

where a is the averaged magnitude of oscillation:

$$a = \frac{y^{max} - y^{min}}{2}$$

and y^{max} and y^{min} stand for the maximum and minimum magnitude of the outputs. Δa is the bias in the asymmetrical oscillation (Figure 7.3B). The value of the *BSR* falls between 0 and 1. If $BSR = 0$, then this means we have a symmetrical oscillation and bias occurs when $BSR > 0$. Qualitatively, we classify the load effect into three levels: low, medium and high. Corresponding errors in K_u and ω_u for first-, second- and third-order plus dead time systems can also be found from Chapter 7. Table 12.1 summarizes the qualitative results.

These are the results obtained using an ideal (on–off) relay. Table 12.1 shows that the *BSR* gives a good indication of the accuracy. For the medium- and high-level load effects, a better experimental design is needed to improve the accuracy.

12.1.3 Nonlinearity

Another important characteristic of industrial processes is nonlinearity. Two types of nonlinearity are addressed. One is the "local" nonlinearity and the other is the "global" nonlinearity. Let us consider the case of "local" nonlinearity. It is obvious that process nonlinearity differs from load disturbance in its origin. But, in a relay feedback test, they show a similar response in the output: asymmetry in the oscillation. Again, we use the *BSR* to describe the degree of nonlinearity. Table 12.2 gives qualitative results derived from a Hammerstein model with different relay heights. Generally, pH neutralization processes, high-purity distillation columns, and reactive distillation belong to the class of medium to high *BSRs*. It also should

Table 12.1. The load effect and corresponding errors

Load level	Range of BSR^*	Error in K_u^{**}	Error in ω_u^{**}
Low	$0 \leq BSR < 0.2$	$< 5\%$	$< 5\%$
Medium	$0.2 \leq BSR < 0.6$	$< 10\%$	$< 20\%$
High	$0.6 \leq BSR < 1$	$< 60\%$	$< 80\%$

* $BSR = \Delta a / a$.
** Use $BSR = 0$ as base value.

Table 12.2. The effect of nonlinearity and corresponding errors

Load level	Range of BSR^*	Error in K_u^{**}	Error in ω_u^{**}
Low	$0 \leq BSR < 0.2$	$< 5\%$	$< 5\%$
Medium	$0.2 \leq BSR < 0.6$	$< 20\%$	$< 20\%$
High	$0.6 \leq BSR < 1$	$< 80\%$	$< 80\%$

* $BSR = \Delta a / a$.
** Use $BSR = 0$ (*i.e.* linear system) as base value.

be noted that, for the same process, different relay heights can lead to different BSRs.

"Global" nonlinearity may arise when the plant is operated over a wide operation range (*e.g.* different throughputs, various product grades). The autotuning can be implemented in a multiple-model framework, as shown in Chapter 8.

12.1.4 Noise

Any practical identification procedure should be able to overcome process and/or measurement noise. For relay feedback tests, noise generally will *not* deteriorate the estimate of K_u and ω_u. However, it may result in apparently random switching of the relay. Therefore, the noise effect may prolong or fail the relay feedback test. To overcome the possible failure, a relay with hysteresis is a good choice. However, a relay with hysteresis means that we are finding a frequency point smaller than the ultimate frequency. In other words, we simply overestimate the magnitude of $|G|$ (getting smaller K_u). Therefore, the errors often come from the way we handle the noise, not from the noise itself.

12.1.5 Imperfect Actuator

The relay feedback may lead to erroneous ultimate properties when we have imperfect actuators, especially for a control valve with actuator. Thus, the relay feedback procedure has to be modified to find the dead band in the hysteresis and the correct ultimate gain and ultimate period. Because imperfect actuators are often encountered in process industries, it is important to find the correct ultimate properties under this circumstance. A two-step procedure is proposed in Chapter 10 to overcome the problem of an imperfect actuator. A final note is that, regardless of how intelligent the procedure is, the best solution is to get the valve fixed.

12.2 Available Relays

In this section we will summarize the relays we have discussed so far. Basically, they can be classified into three types.

1. *Ideal (on–off) relay*
   As mentioned several times, this is simplest relay. The disadvantage is that the estimated K_u and ω_u are less accurate. But, it provides a simple and reliable way for autotuning.

$$K_u = \frac{4h}{\pi a} \quad (12.2)$$

$$\omega_u = \frac{2\pi}{P_u} \quad (12.3)$$

2. *Saturation relay*
   The saturation relay compensates for the problems associated with the ideal relay. However, it requires a longer test period due to changes in the slope k. Initially, the slope of a saturation relay is set to a high value (e.g. $k \to \infty$ implies an ideal relay). Then, we can have a rough estimate of the ultimate gain $\hat{K}_u$. Next, the desired slope is calculated from

$$\begin{aligned} k &= 1.4 k_{min} \\ &= 1.4 \hat{K}_u \end{aligned} \quad (12.4)$$

The next step is to continue the test with this slope and a better estimate of K_u can be found:

$$K_u = \frac{2h}{\pi \bar{a}} \left[\left(\sin^{-1} \frac{\bar{a}}{a} \right) + \frac{\bar{a}}{a} \sqrt{1 - \left(\frac{\bar{a}}{a} \right)^2} \right] \quad (12.5)$$

3. *Biased relay*
   (i) Output-biased relay

As pointed out in Chapter 7, the output-biased relay is effective in overcoming asymmetrical oscillation. The reason for restoring the symmetry is that the asymmetry gives rise to significant errors in K_u and ω_u. Asymmetrical responses often arise from load disturbances and process nonlinearity. The biased value of the relay is adjusted on-line according to

$$\delta_0^{(i)} = \delta_0^{(i-1)} - \frac{h \Delta a^{(i)}}{a^{(i)}} \tag{12.6}$$

where δ_0 is the bias in the relay height, as shown in Figure 7.5A.

(ii) Input-biased relay

Input-biased relay is useful in generating asymmetrical oscillation. In doing this, we can find the steady state gain K_p, in addition to the K_u and ω_u, from a single relay feedback test [4]. According to the equivalent gain in the describing function, K_p can be expressed as

$$K_p = -\frac{\int_0^{2\pi} e(t) \, d\omega t}{\int_0^{2\pi} u(t) \, d\omega t} \tag{12.7}$$

Therefore, in the normal operating condition (e.g. without load disturbance and severe nonlinearity), we can obtain both steady state (K_p) and dynamic information (K_u and ω_u) using the input-biased relay feedback.

12.3 Specifications

Depending on the application, the relay feedback experiment can be designed for specific purposes. Two typical controller design methods are direct tuning and model-based tuning (Chapter 2). From the process perspective, the applications can also be classified as single-loop and multiloop systems. In this section, we will look at the direct tuning, model-based tuning and multiloop applications.

12.3.1 Direct Tuning

The process information required for direct tuning (Table 2.2) is the ultimate gain K_u and the ultimate frequency ω_u. This is the situation where simple relay feedback tests work well. Table 12.3 gives the appropriate relays for different process conditions.

In normal operating conditions, the saturation relay will improve the accuracy of the identification. Under load changes or facing process nonlinearity, the output-biased relay is recommended, since it reduces the identification error significantly. Table 12.3 probably covers most of the process applications. A pH neutralization example is used to illustrate the effectiveness of the output-biased relay.

Table 12.3. Available relays for direct tuning

Operating condition		Available relay
Normal		or
Measurement noise (low, medium, high)		or
Load disturbance	low	or
	medium	
	high	
Nonlinearity	low	or
	medium	
	high	

Example 12.1 pH neutralization process.
This is a strong nonlinear process where the strong base NaOH is employed to neutralize waste acids: HCl and HAc.

First an ideal relay with a height of 10% (F_{base}) is used to generate continuous cycling (t = 0–4 min in Figure 12.1). Output responses show that the oscillation is asymmetrical ($a = 0.16926$ and $\Delta a = 0.0353$). We can find the ultimate gain and ultimate frequency from Figure 12.1. They are $K_u = 0.01599$ and $\omega_u = 5.81$. K_u is off by 5.2%, if we use a relay height of 0.1% (can only be done in simulation). According to Equation 12.6, an output bias of 20% (δ_o) is established and another relay feedback test is performed. Figure 12.1 shows that symmetrical oscillation is restored and K_u and ω_u become: 0.01692 and 7.75 respectively. The error in K_u is less than 1% using the output-biased relay. Note that this is achieved with a relay height of 10%. This example clearly illustrates the effectiveness of the output-biased relay. ∎

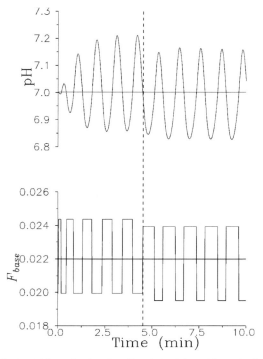

Figure 12.1. Output-biased relay feedback test for pH neutralization system

12.3.2 Model-based Tuning

In some cases we would like to try out model-based tuning. IMC-PID tuning is a typical example where we need to have a linear model for the controller tuning (Table 2.3). Despite the fact that we can find a linear model from an ideal relay feedback (*e.g.* Chapter 3), a more reliable approach is to find the steady state gain and ultimate gain and frequency from the relay feedback tests. Therefore, for a simple FOPDT model, the required information is K_p, K_u and ω_u. Table 12.4 shows the available relays for this situation. Here, we recommend two consecutive relay feedback tests to obtain all three items of process data. For example, in the normal operating condition, an input-biased relay feedback is performed first to find the K_p, followed by a saturation relay to obtain K_u and ω_u. In doing this, we can have a better estimate of three process values. Table 12.4 also indicates that, in theory, we can identify these three values under load disturbance. That is, we can use an output-biased relay to find K_u and ω_u followed by an input-biased relay to get K_p. But, this is generally not recommended in practice, since it is rather sensitive to process changes, *i.e.* non-stationary load disturbances.

Table 12.4. Available relays for model based tuning

Operating condition	Available relay
Normal	
Measurement noise	
Load disturbance (low, medium, high)	
Nonlinearity (low, medium, high)	

Let us use the WB column example to illustrate the two-stage relay feedback tests.

Example 12.2 WB column.

$$G(s) = \frac{12.8e^{-s}}{16.7s+1}$$

First, an input biased relay with $\delta_i = 0.1$ is used to generate a sustained oscillation (t = 0–14.5 min in Figure 12.2). The results are $K_p = 12.792$, $K_u = 1.72$ and $\omega_u = 1.616$. In spite of good accuracy in K_p, the ultimate gain is off by almost 20%. Next, the slope of the saturation relay is calculated ($k = 1.72 \times 1.4 = 2.41$) and a second relay feedback test is carried out (t = 14.5–30 min). The results show that K_u and ω_u become 2.098 and 1.606, respectively. The corresponding errors are −0.01% and 0.012% respectively. Obviously, the two-stage approach is quite effective in finding all three process data. ∎

12.3.3 Multiloop System

We have discussed multiloop autotuning in Chapter 6. With the sequential identification and tuning procedure, we are able to obtain an interaction measure (*i.e.* RGA [5]) using the multiloop version of the input-biased relay. Since we can obtain K_p using the input-biased relay, it is possible to construct the gain matrix under the framework of sequential identification. Consider an $n \times n$ multivariable system. From the first n autotuning steps we can calculate all n^2 steady state gains ($K_{p,ij}$, $i,j = 1-n$, or $G(0)$). Once $G(0)$ is available, the RGA can be found accordingly [5].

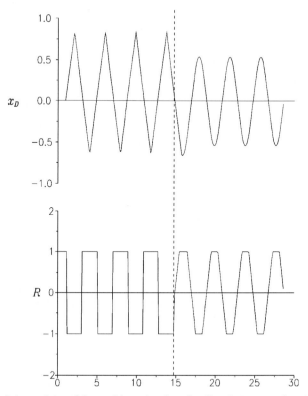

Figure 12.2. Integration of input-biased relay feedback test and saturation relay feedback test for WB column

$$RGA = \mathbf{G}(0) \otimes \left(\mathbf{G}(0)^{-1}\right)^T \qquad (12.8)$$

where $\otimes$ stands for element-by-element multiplication and the superscript T denotes the matrix transpose. It is useful to obtain the interaction measure along the tuning steps. More importantly, it can be used to re-evaluate the appropriateness of the variable pairing. However, it is important to ensure the consistency of the steady state gains. The sequential identification procedure of Shen and Yu [4] is useful in this regard.

Let us use a 2×2 example to illustrate these sequential identification steps. Starting from the first loop, we have

$$K_{p,i1} = \frac{\int_0^{2\pi} y_i(t)\, d\omega t}{\int_0^{2\pi} u_1(t)\, d\omega t}, \quad i = 1, 2 \qquad (12.9)$$

This step is exactly the same as the conventional identification procedure. Next, perform an input-biased relay feedback test while loop 1 is on automatic. The entry in the second column of the gain matrix becomes

$$K_{p,i2} = \frac{\int_0^{2\pi} y_i(t)\, d\omega t - K_{p,i1} \int_0^{2\pi} u_1(t)\, d\omega t}{\int_0^{2\pi} u_2(t)\, d\omega t}, \quad i = 1, 2 \quad (12.10)$$

Equation 12.10 can be viewed as solving the following simultaneous equations:

$$y_1 = K_{p,11} u_1 + K_{p,12} u_2 \quad (12.11)$$

$$y_2 = K_{p,21} u_1 + K_{p,22} u_2 \quad (12.12)$$

while $K_{p,i1}$ is known. The unknown $K_{p,i2}$ then becomes

$$K_{p,i2} = \frac{y_i - K_{p,i1} u_1}{u_2}, \quad i = 1, 2 \quad (12.13)$$

Therefore, we can repeat the multiloop autotuning procedure in Chapter 6 (Figure 6.16) with the input-biased relay. For a general $n \times n$ system, the equation for the steady state gains under sequential identification is

$$K_{p,in} = \frac{\int_0^{2\pi} y_n(t)\, d\omega t - \sum_{j=1}^{n-1} \left(K_{p,ij} \int_0^{2\pi} u_j(t)\, d\omega t \right)}{\int_0^{2\pi} u_n(t)\, d\omega t}, \quad i = 1 - n \quad (12.14)$$

After n steps, we obtain the $\mathbf{G}(0)$ and the RGA can be evaluated using Equation 12.8. Decisions on continuing tuning or control structure reconfiguration can be made with the interaction measure. Several authors have discussed the implications of the RGA [6–9]. The blending example of Chapter 6 is used to illustrate the enhanced multiloop autotuning.

Example 12.3 Blending system.
Consider a blending system (Figure 6.9) with two feed streams F_1 and F_2. The control objective is to maintain the outlet flow rate F and concentration x by manipulating F_1 and F_2. At nominal condition we have $x = 0.78$, $F = 20$, $x_1 = 0.3$, $F_1 = 4$, $x_2 = 0.9$ and $F_2 = 16$. In a nonlinear simulation, 0.5 min of measurement delay is added to the composition measurement.

Initially, the control system is set up as follows. The total flow is controlled by F_1 and the concentration is maintained by changing F_2. We proceed with the multiloop autotuning using an input-biased relay. First, a relay feedback test is carried out on the $F - F_1$ loop with $h = 1$ and $\delta_i = 0.1$ ($t = 0$–2.35 in Figure 12.3).

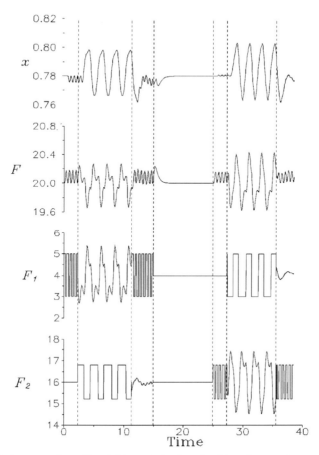

Figure 12.3. Improved multivariable autotuning for blending system with different variable pairings

From Equation 12.9 we obtain the steady state gain ($K_{xF_1} = -0.0238$ and $K_{FF_1} = 1$) and controller parameters are $K_c = 4.67$ and $\tau_I = 1.06$. When the $F-F_1$ loop is on auto mode, a second input-biased relay feedback test is performed on the $x-F_2$ loop. From system responses (t = 2.35–11.3 min in Figure 12.3), we have $K_{xF_2} = 0.0063$ and $K_{FF_2} = 1$. The controller parameters from the Shen–Yu formula become $K_c = 21.2$ and $\tau_I = 3.74$. We can calculate the RGA immediately. This gives $\lambda_{11} = 0.21$. This value is very close to the true value ($\lambda_{11} = 0.2$). Moreover, it indicates that the other pairing ($x-F_1$ and $F-F_2$) is more appropriate, since the RGA is close to 1 ($\lambda_{11} = 1 - 0.21 = 0.79$). If we choose to reconfigure the control structure (*i.e.* $x-F_1$ and $F-F_2$), then the controllers have to be redesigned. The autotuning procedure is shown in Figure 12.3 (t = 25–38 min). Controller settings for the $x-F_1$ loop and $F-F_2$ loop are $K_{c1} = -21.7$, $\tau_{I1} = 4.69$ and $K_{c2} = 3.35$, $\tau_{I2} = 1.10$.

Figure 12.4 clearly shows that the reconfigured control structure is less interacting, since the RGA is close to 1. The example reveals the potential benefit of the input-biased relay: obtaining the interaction measure along with controller parameters. ∎

12.4 Discussion

It is always desirable to use the right tool in the right way, at the right time. Controller tuning is no exception to this. Process understanding is essential, so that the current situation can be correctly assessed and the appropriate tool can be applied accordingly. The controller cannot function well if the model is not correctly identified. Thus, the importance of the experimental design in system identification cannot be overlooked.

A summary, based on the discussion and observations, of the available relays for various operating conditions is given in Table 12.5.

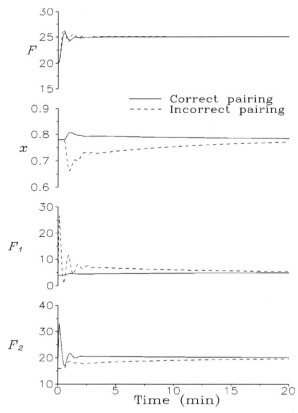

Figure 12.4. Step SP responses for blending system with two different variable pairings

Table 12.5. Summary of available relays for different operating conditions

Operating conditions	Information requirement	Available relay
Normal	K_u and ω_u	(relay) or (relay)
	K_p, K_u and ω_u	(relay) or (relay)
Measurement noise	K_u and ω_u	(relay) or (relay)
	K_p, K_u and ω_u	(relay) or (relay)
Load disturbance	K_u and ω_u	(relay)
	K_p, K_u and ω_u	(relay)
Nonlinearity	K_u and ω_u	(relay)
	K_p, K_u and ω_u	(relay)

12.5 Conclusion

Accurate data is the key to good control performance. In this chapter we have tried to integrate various relays with familiar operating conditions. The real difficulty in autotuning is not in the methodology itself, it is in the selection of an appropriate tool for the right situation. The intelligence of plant operators should exceed that of the commercial intelligent controller. In terms of the relay feedback test, the rules are:

1. Use an ideal relay feedback test to find a rough estimate of K_u and ω_u and to determine corresponding operating condition.
2. Use the saturation relay to improve accuracy.

3. Use a relay with hysteresis to overcome noises, but you should recognize the possibility of underestimating K_u.
4. Use an output-biased relay to restore symmetrical oscillation if the original cycling is not symmetrical.
5. Use an input-biased relay to obtain steady state gains if they are really necessary.

In terms of controller design, we recommend:

1. Use the "shape" of the relay feedback response to identify model structure and to find the appropriate tuning rule.
2. Use "sequential" design to handle multivariable processes.
3. Be aware that erroneous estimates of the ultimate gain and ultimate period may arise when the control valve shows a dead band and (or) hysteresis. The two-step procedure in Chapter 10 should be taken to restore the correct information.
4. Integrate "autotuning" into the multiple-model framework to handle global nonlinearity.

12.6 References

1. Åström KJ, McAvoy TJ. Intelligent control. J. Process Cont. 1992;2:115.
2. Åström KJ, Hang CC, Persson P, Ho WK. Towards intelligent PID control. Automatica 1992;28:1.
3. Åström KJ, Hägglund T. The future of PID control. Cont. Eng. Prac. 2001;9:1163.
4. Shen SH, Yu CC. Sequential identification: A procedure for insuring internal consistency of steady state gains for multivariable systems. Can. J. Chem. Eng. 1996;74:132.
5. Bristol EH. New measure of interaction for multivariable process control. IEEE Trans. Auto. Control 1966;AC-11:133.
6. McAvoy TJ. Interaction analysis. Instrument Society of America: Research Triangle Park; 1983.
7. Grosdidier P, Morari M, Holt RB. Closed-loop properties from steady state gain information. Ind. Eng. Chem. Process Des. Dev. 1985;24:221.
8. Skogestad S, Morari M. Implications of large RGA elements on control performance. Ind. Eng. Chem. Res. 1987;26:2323.
9. Yu CC, Fan MKH. Decentralized integral controllability and D-stability. Chem. Eng. Sci. 1990;45:3299.

Index

ATV, 23
averaging level control, 233

balanced scheme, 227
bias value
 load disturbance, 144
 output-biased relay, 146
bias-to-signal ratio, 246
blending system, 109, 254
bump test, 202

C_3 splitter, 124
Ciancone–Marlin tuning, 18
closed-loop characteristic equation, 103
closed-loop transfer function, 91, 102
complementary sensitivity function, 103, 113
control configuration, 1
control structure design, 222
controller gain, 9
controller structure, 1
controller tuning, 1
 entire plant, 232
conventional scheme, 232
convergence, 117
critical slope, 83
CSTR, 220

dead band, 199
dead time, 34
 compensation, 1, 70
 error, 36
dead-zone, 197
derivative kick, 16
derivative time, 12

describing function, 77, 81
distillation column, 124
disturbance sensitivity, 231
dual input describing function, 139

equivalent gain, 139, 249
experimental design, 24, 84

first-order plus dead time, 17, 27, 32, 48, 52, 55, 62, 189
Fourier transformation, 31, 76, 80, 141
fuzzy model, 157, 161, 164, 170

gain margin, 163
Gauss–Seidel method, 117
global model, 157, 162, 170

high-order system, 52, 58, 59, 64, 71, 194
high-purity distillation column, 23, 87, 124
hysteresis, 197, 205
 width of the hysteresis, 199, 202, 205, 208, 211, 212

identification
 independent, 108
 sequential, 108
IMC-PID tuning, 19, 63, 65, 251
imperfect actuator, 197, 206, 213, 248
instrumentation, 1
integral controllability, 2
integrated absolute error, 61, 62, 179
integrator plus dead time, 27, 156, 162, 177
integrity, 121, 229

intelligent control, 245
internal model control, 2, 34
inventory control, 233
inverse response, 235
ITAE, 63

load disturbance, 135, 138, 210, 246
local model, 156, 164, 170
Luyben structure, 226

maximum closed loop log modulus, 64, 113, 175
M-circle criterion, 98
measurement noise, 67, 68
MIMO, 90, 93, 97, 99, 252
model predictive control, 3
moderate-purity distillation column, 90, 91, 92, 152, 220
modified z-transform, 30

noise, 247
noise-to-signal ratio, 67, 207
nonlinearity, 246
normalized dead time, 25
normalized loop speed, 119
Nyquist plot, 111

odd-symmetric function, 77
overdamped, 39

P control, 9
parameter estimation, 32
perfect control, 103
performance assessment, 178
pH neutralization, 249
phase margin, 64, 163
PI control, 10
PID control, 12, 62
PID controller, 3
    ideal PID, 12
    IMC PID, 15
    parallel PID, 12
    series PID, 15
plantwide control, 219
pole
    RHP, 132
    sequential design, 105
procedure, 89, 115, 122, 148
process design, 2
production rate change, 239
proportional kick, 16
pseudo-random binary signal, 4

pulse test, 5

quality control, 235

ramp test, 202, 212
rangeability, 231
ratio control, 233
reactor/separator, 212
recycle system, 222
relative gain
    negative, 107
relative gain array, 2, 103, 228, 252, 254
relay
    biased, 44, 248
    ideal, 24, 75, 148, 197, 200, 202, 205, 248
    input-biased, 249, 252, 254
    output-biased, 142, 149, 248
    saturation, 78, 83, 197, 199, 200, 202, 248
relay feedback, 4, 23, 25
    load disturbance, 137, 144
reset time, 11
reset windup, 13
robust performance, 163
robust stability, 163
R–V control structure, 90, 152

second-order plus dead time, 27, 34, 56, 64, 136, 189
sequential design, 99, 101
sequential tuning, 236
shape, 47, 245
    relay feedback, 176
Shen–Yu tuning, 111, 212, 249
singular value tuning, 127
SISO, 97
snowball effect, 222
spectral radius, 118
steady-state gain, 26, 249, 252, 255
step test, 4, 25
    load disturbance, 136
stepping technique, 152

T4 column, 128
Takagi–Sugeno model, 157, 161
Tennessee Eastman process, 160, 167, 170
third-order plus dead time, 149
transmitter span, 215
tuning sequence, 119
Tyreus–Luyben tuning, 18, 62, 68

ultimate frequency, 24, 137
ultimate gain, 24, 137, 197, 199, 205, 210
ultimate period, 24, 30, 199, 205
unbalanced scheme, 222
underdamped, 105
unstable system, 53, 59, 63

variable pairing, 120

WB column, 29, 84, 89, 105, 116, 252

zero, 107
  RHP, 88, 107
  sequential design, 105
Ziegler–Nichols tuning, 3, 17, 111, 199, 206

Printed by Publishers' Graphics LLC